Farmer Organizations: Status and Prospects

Farmer Organizations: Status and Prospects

S Ayyappan
Letha Devi G
Sreenath Dixit

SSPH

SATISH SERIAL PUBLISHING HOUSE
403, Express Tower, Commercial Complex
Azadpur, Delhi-110033 (India)
Phone : 011-27672852, Fax : 91-11-27672046
e-mail : info@satishserial.com, hkjain1975@yahoo.com
Website : www.satishserial.com

Published by :

SATISH SERIAL PUBLISHING HOUSE
403, Express Tower, Commercial Complex, Azadpur, Delhi-110033 (INDIA)
Phone : 011-27672852 Fax : 91-11-27672046
E-mail : info@satishserial.com, hkjain1975@yahoo.com

₹ 1150/-

ISBN 978-93-86200-77-8

Composed, Designed & Printed in India

Preface

Agriculture is the most discussed topic today across the country, for the food production, for the livelihoods, for the export earnings, and for the confidence it has brought to the country, as also numerous issues. The contextual framework comprises the challenges of climate change, natural resources degradation, emerging diseases and pests, labour issues, post harvest losses, rural to urban migration and so on. Expectations are those of higher productivity and profitability, doubling farmers' income, skilling agriculture and making farming an enterprise, keeping in mind the monsoon, markets and the mindsets.

Integration, diversification, clustering, input management, farm mechanisation and on-farm value addition are being proposed in this value chain approach, from seed to market. At this stage, Farmer Organisations are being perceived as a potential mechanism to contribute and enhance farmers' income. SHGs, FIGs, FPOs and FPCs are all in different various stages of evolution. Lessons learnt from informal groupings to corporates, overseas experiences from Africa to Europe, are being sourced to realise the aspirations of the Farmer Organisations.

With several initiatives by the NABARD, SFAC, Governmental Departments, and the National Policy for FOs, 2013, the approach has been to bring professionalism into FOs, for according environmental and ecological health, knowledge and understanding of local farming practices, achieve sustainability and enhance farmers' income. In the last two decades, among different forms of FOs, there have been several success stories, clarity of concept and visionary approach, organisational coherence, entrepreneurial capacity demonstration, market intelligence and hard work of the members to achieve the goals. These have also brought out the importance of the operating environment, from advisories to input management to value addition to sale of produce. At the core of the success are, group coherence, financial health and entrepreneurship.

In this background, the NABARD Chair, ICAR-ATARI, NDRI, NIANP and ISEE (KC) organised a National Consultation on 'Farmer Organisations: Status & Prospects' at Bengaluru during 25-26 July, 2017. Over 150 Professionals, including Experts and Farmers participated in the event, with lead talks, panel and group discussions, encompassing all forms of FOs. The deliberations brought out specific action points regarding Capacity building, Resource use efficiency, Management,

Marketing & Value Chain and Policy support, providing a Way forward for empowering FOs for remunerative agriculture.

We are grateful to Dr H.K. Bhanwala, Chairman, NABARD, Mumbai; Dr H.R. Dave, Deputy Managing Director, NABARD, Mumbai; Dr P.M. Ghole, Chief General Manager, NABARD, Mumbai; Dr U.S. Saha, Chief General Manager, NABARD, Mumbai; Dr Debesh Roy, Deputy General Manager, NABARD, Mumbai; Dr M.I. Ganagi, Chief General Manager, NABARD, Bengaluru; Dr B.S. Suran, Managing Director, NABFINS, Bengaluru; Dr Raghavendra Bhatta, Director, NIANP, Bengaluru; Dr K. Narayana Gowda, President, ISEE (Karnataka Chapter); Dr K.P. Ramesha, Head, NDRI-SRS, Bengaluru, for the guidance and facilitations of the National Consultation. Thanks are due to the Lead Speakers, Delegates and Participants for enriching the Proceedings of the Conference.

1 January, 2018 **Editors**

About the Editors

Dr S Ayyappan, has significantly contributed to the R&D of fisheries sector in particular and agriculture in general. He has served the country in different roles of Director, ICAR-CIFE, Mumbai; DDG (Fisheries), ICAR, New Delhi and Secretary, DARE and Director General, ICAR before joining as the NABARD Chair Professor at ICAR-NDRI, SRS, Bengaluru. His scientific contributions towards the developments of agriculture are well appreciated and he has been bestowed with many patents and awards.

Dr Letha Devi G is currently working as Senior Scientist at ICAR-NIANP, Bengaluru. Her research interests include climate change and Indian livestock and ICTs in agriculture. She has been associated with various research projects on climate vulnerability analysis of livestock farming and development of various ICT tools for livestock farmers.

Dr Sreenath Dixit is currently the Director of ICAR -ATARI, Bengaluru. Earlier he worked at ICAR-NRC Cashew, Puttur; ICAR-CRIDA, Hyderabad and ICRISAT, Hyderabad. He has successfully steered several projects funded by national and international agencies like the DFID, ACIAR and the World Bank. He has over 180 publications to his credit. He has won several awards including Vasantrao Naik Award for dryland agriculture and Fakhruddin Ali Ahmed Award for Tribal Agriculture improvement and Young Scientist award of the Indian Society of Extension Education early in his career.

Contents

Chapter 1

Farmer Producer Organisations: NABARD's Perspective

Ganagi M I

NABARD, Karnataka Regional Office, Bangalore

It is well recognized that fast agriculture growth is the most effective route for inclusive growth, especially when the share of agriculture in Gross Value Added (GVA) is around 15 per cent, (2015-16) and shared by more than half of our population. Simple mathematics shows that the per capita income of those involved in agriculture is almost one third of an average Indian. India had over 138 million farm holdings as per the Agricultural Census, 2011. Of this, about 92.8 million were marginal farm holdings i.e. having individual operational land holding of less than 1 hectare while about 24.8 million were small farm holdings with individual operational land holding size of less than 2 hectares. Therefore, the marginal and small farm holdings together accounted for a whopping 85 percent of the total farm holdings in India in 2010-11. Hence, every effort to inclusive growth has to address the income enhancement in agriculture and particularly the weak and vulnerable members within the sector.

Landlessness, fragmentation and smallness of operations is not a new phenomenon, but the stage at which it stands now, has serious consequences in almost every aspect of agriculture growth and development - be it production, storage, transporting, marketing and most importantly, incomes.

Further the declining of average size of land has implications for agriculture credit outreach too. Ironically, it is also true that small and marginal farmers account for a substantial share in vegetables and milk, cereals and fruits produced. But, with very little marketable surplus, their farming is hardly commercial. Due to continued fragmentation of landholding, rising cost of cultivation, lack of assured market, modern technologies and quality inputs, generally the small/

marginal land holders are not able to realize optimal price for their produce and productivity continues to be sub-optimal.

As the small farms struggle to get access to inputs, market and credit – they need a level playing field to be able to compete with other market players in equal terms. Due to factors beyond their control and absence of institutions to safeguard their interests, they are unable to integrate with the agricultural value chains, fight the risks and vulnerabilities such as commodity price volatility, crop failure, insect pest-attacks etc. on their own. Considering the legal environment and political sensitivity of the agricultural sector, there is a gradual recognition that one of the ways to overcome the challenges as described above, could be to collectivize the farmers into Producer Organizations (POs).

Collectivization of producers, especially small and marginal farmers, into producer organisations has emerged as one of the most effective pathways to address the many challenges of agriculture but most importantly, improved access to investments, technology and inputs and markets. Department of Agriculture and Cooperation, Ministry of Agriculture, Govt. of India has identified farmer producer organisation registered under the special provisions of the Companies Act, 1956 (as amended in 2013) as the most appropriate institutional form around which to mobilize farmers and build their capacity to collectively leverage their production and marketing strength.

Producer Organisations

A Producer Organization (PO) is a legal entity formed by primary producers like farmers, milk producers, fishermen, weavers, rural artisans, craftsmen, etc. A Producer Organization formed by the farmers or farmer organizations, is known as Farmer Producer Organization (FPO). The advantages envisaged in Producer Organization (PO) are of a group of producers for either farm or non-farm activities, a registered body and a legal entity, producers are shareholders in the organization and deals with business activities related to the primary produce/products, works for the benefit of the member producers, a part of the profit is shared amongst the producers and rest of the surplus is added to its owned funds for business expansion, the ownership of the PO rests with its members and it is an organization of the producers, by the producers and for the producers.

The FPOs could be facilitated by an experienced person/ NGO/ Community Based Organization, Corporates, Banks, Financial Institutions, Development Agencies, Government Departments, KVKs, Commodity Boards, Agri Universities/research Agencies, Cooperative Societies, Large Producer Companies, Producer Associations or Federations. It could take off through different legal forms viz., Cooperative Societies Act/Autonomous or Mutually-aided Cooperative Societies Act of the respective State, Multi-State Cooperative Society Act, 2002, and Producer Company under Section 581(C) of Indian Companies Act, 1956, as amended in 2013.

The key elements that could be kept in view while forming POs are current and future demand of produce in the existing market, willingness of producers to invest and adopt new technology and undertake collective action, accept challenges in the market chain and market environment, vulnerability of the market to shocks, trends and seasonality, key commodities, processed products or semi-finished goods demanded by major retailers or processing companies in the surrounding areas/districts etc.

In this scenario, there are around 3500 FPOs in the country, which are promoted by SFAC, NABARD, State Govt. departments, NGOs and other promoting organisations.

The major benefits to farmers by organizing into an FPO are :

1. **Increased scale of operation:** POs bring together the business activities of small-scale producers/farmers and thereby increase the overall scale of operation, which, in turn, lowers the cost of inputs and business transaction due to economies of scale. On account of aggregation, the PO is able to maintain consistency in the supply of agricultural produce both, in terms of quantity and quality.
2. **Improved market access:** Collectivization/aggregation of agricultural produce enables small-scale producers to build adequate scale as per the needs of the buyers. Also, processing of large quantity of produce enables PO to access higher value markets at a later stage and realise better price for their produce.
3. **Increased bargaining power:** Collective action can increase the bargaining power particularly of small/ marginal producers in various ways. Economies of scale can make it easier for a PO to invest in processing, storage or transport facilities that can give

members the ability to choose when and where to sell their produce.

4. **Improved Access to Services:** Firstly, the lower transaction costs and economies of scale involved in providing services to groups rather than to individual producers, makes it much cheaper and easier for businesses and service providers to work with small-scale producers. Secondly, organized and legally registered groups of producers, are likely to have greater credibility and a higher profile than individual producers.

I. NABARD's initiatives in the development of FPOs

1. Promotion of FPOs under PRODUCE Fund

NABARD has been promoting new FPOs out of a dedicated PRODUCE Fund created by the Govt. of India during 2014- 15. Under this fund, over 2100 FPOs have been formed with a grant commitment of Rs.193.90 crore under various legal statutes across 29 states. The grant assistance sanctioned per FPO was to the tune of Rs. 9.06 lakhs. These FPOs are under various stages of formation, with 1924 FPOs having completed their registration formalities by 31 March 2017. Out of these FPOs, 149 were registered under Companies act (as amended in 2013), and the remaining FPOs were registered under Cooperative Societies Act/Trust Act. As on 31 march 2017, cumulative share capital collected by FPOs stood at Rs. 41.60 crore and around 7 lakh farmers were covered. The major activities in which FPOs are engaged include production and marketing of commodities like cereals, pulses, spices, jiggery, organic inputs/processing, onion, fruits, vegetables, certified seeds, etc. majority of FPOs are undertaking farm aggregation, marketing and input distribution activities. About 1515 FPOs are undertaking business activities, and their business turnover is about Rs. 100 crore.

2. Credit plus support to POs

NABARD continued to provide credit cum grant support to POs out of PODF up to September 2016 for meeting their investment needs and also working capital requirements. However, since direct loan to POs from NABARD has been discontinued from October 2016, NABKISAN Finance Ltd (NKFL), a subsidiary of NABARD has been assigned the task of providing adequate credit support to POs on affordable terms. So far, NABKISAN Finance Ltd., has supported 210

FPOs across the states and the cumulative sanctioned amount is 71.55 crores with a disbursement of Rs. 60.46 crores.

3. Demand for promotion of more FPOs

In view of the current policy focus, there has been growing demand from various quarters and farmer group, for organisation of more number of FPOs due its huge potential and requirement of such collectives particularly in uncovered areas/regions where there is large proportion of SF/MF. The successful experiences and emerging opportunities in terms of better prospects for marketing of produce and realisation of optimal price for the produce are inspiring other farmers to come forward and get organised into suitable collectives. It was therefore proposed to promote additional 3000 FPOs on a cluster basis or in a compact area keeping in view marketing opportunities and infrastructure from the operational point of view as also economise the cost of monitoring, training, handholding and overall management. The formative expenses for promotion of these new FPOs are to be met out of the promotional funds of NABARD on a project-to-project basis subject to a ceiling of Rs. 2.00 lakhs per FPO for a two-year period inclusive of a maximum of Rs.40000/-towards registration charges.

4. Development of FPOs Rating Tool

Amidst the spur in promotion of the FPOs over the last decade, there is a need to differentiate the well-functioning and credit worthy business entities/FPOs. On the lines of the rating tool for the microfinance institutions, NABARD has developed a rating mechanism, which will be done periodically to rate the functioning of the FPOs on activities such as governance, management, risks mitigation, and sustainability.

5. Digitisation of FPOs

NABARD had launched a pilot in Karnataka for digitising the entire process of FPOs for on-line monitoring and planning. The web portal would capture key data on FPOs, including business transaction details entire KYC details of members, updation on real-time basis, all processes/ activities of FPOs with Geo coding and Geo tagging features, alert mechanism for submission of returns and other compliances, creation of templates on Analytics to facilitate effective monitoring by Stakeholders.

II. Initiatives by SFAC

Small Farmers' Agri-Business Consortium (SFAC) was mandated by Ministry of Agriculture, Government of India to lead a national pilot project to promote POs as a demonstration of the benefits of building institutions of producers and their integration in agri-value chains. Starting in 2011, SFAC has so far set up nearly 250 FPOs covering 2.85 lakh farmers spread across 25 states, most of them being registered under Companies Act, 2013.

The typical PO promoted by SFAC usually has 15-20 farmers in a group (referred to as a Farmers Interest Group (FIG)) and 50 FIGs covering around 1000 farmers will form a PO at the cluster level. SFAC hopes to federate POs at the state level at a later stage. In 2014, DAC has set up the National Policy and Process Guidelines for Farmer Producer Organizations, along with providing various other policy reforms in credit and procurement for SFAC to implement. SFAC is therefore a critical institution to promoting POs in the country, and ensuring that their structure and functioning is beneficial (and suitable) to their corresponding roles within states.

Potential for new FPOs

As on date, around 3500 FPOs are existing in the country with cumulative membership of around 8.75 lakh, of which 70-80% members are small and marginal farmers. Presently there are around 1.48 lakh FCs, 637 wadi/orchard development and 1607 watershed projects, involving around 21 lakh farmers, most of them are small and marginal farmers. In addition, about 93000 PACS are in existence with a cumulative borrowing membership of around 5 crore members.

Considering the large number of farmers already associated with various informal collectives, there is a vast potential for formation/ promotion of new FPOs through federating these farmer collectives/ farmer clubs/farmers interest groups. Further, out of 12 crore small and marginal farmers in the country, the number of farmers covered so far under FPO segment is very insignificant.

Issues and Challenges

1. **Size of Membership:** The average membership of FPOs is low at around 300. Ideally a membership of 500 to 1000 members in an FPO would ensure sustainability.

2. **Equity mobilization:** A minimum of Rs. 5.00 lakhs may be mobilized by ensuring that all farmers subscribe to share capital. Adequate capital would enable the FPOs take up various business activities. External capital infusion can also help in strengthening the equity base of FPOs.

3. **Lack of professional management:** Lack of professionals has been one of the key limiting factors in the formation and promotion of viable FPOs. In order to develop a cadre of local Professionals, a Short Term Certificate Course (STCC) of 3 months duration was designed by NABARD in consultation with Bankers Institute of Rural Development, Lucknow to provide intensive training to CEO of FPOs formed under PRODUCE Fund.

4. **Delay in taking up business activities:** FPOs should be encouraged and guided to initiate business activities at the earliest to ensure flow of income and thereby viability.

5. **Initial working capital funding without collateral:** Some FPOs due to poor financial capacity/asset base are finding it difficult to access institutional finance for meeting working capital needs. Financial institutions should come forward to extend credit support to FPOs.

6. **Social capital formation for FPO:**The social capital of a society includes the institutions, the relationships, the attitudes and values that govern interactions among people and contributes to economic and social development.

7. **Lack of infrastructure:** Creation of common Infrastructure/ Service Centres to facilitate value addition of produce is essential. The Agri Marketing and Horticulture Department can play a major role in facilitating these services.

8. **Delay in Registration, obtaining Licenses:** Apart from input licenses, FPOs have to obtain various other licenses to carry out business operations. Modalities for simplifying the process of issue of license and completion of registration should be worked out to avoid unnecessary delay affecting the business activities of FPOs. FPOs need to be registered under the Company Act. There is a need to understand the nitty- gritty of company registration act by the farmers and the extension functionaries as facilitators, have a major role to play.

9. **Difficulty in understanding and complying with the provisions of the Companies Act:** It is of extreme importance that the POPI/ FPO realize the roles and responsibilities, terms and conditions, principles of governance etc. It is advisable to appoint an accountant & CEO & possibly a farmer director to be involved in the process from the beginning.
10. **Technical support:** In this tech-savvy world, minimal presence of ICT tools viz., computer, laptop, printer and internet connection in the office of FPOs is desirable.
11. **Market access:** Small and marginal farmers at individual level will not be able to negotiate the price in the market whereas at collective level they can bargain for better prices. Small and marginal farmers through FPOs may access local, provincial, national and international markets. For accruing this benefit, market information, transport and storage facilities, grading, packing etc. are very essential.
12. **Primary value addition at source**: packing, transport, logistic and market- Strengthening agricultural value chains helps in improving efficiency, minimizing postharvest losses, commercialization, improving income and reducing poverty.

Financing FPOs

The FPOs can meet their financial requirement in the following ways :

1. **Share Capital**: The share capital collected from members should be such that it meets the financial needs of initial business activities. As the business grows, share capital can be further strengthened by retaining a small portion of members' profit towards share money as also by enrolling new members.
2. **Grant Support towards Promotional Costs**: Consequent upon amendments in the Companies Act many Govt. organizations like SFAC & NABARD and other development agencies, NGOs/ Corporates, etc., are promoting large number of FPOs by extending financial support (grant or soft loan or both).
3. **Working Capital finance by Banks/NBFCs/MFIs:** AsperRBI circular, financing to producer companies/ cooperatives by banks up to loan amount of Rs. 200 lakh per borrower, would qualify for achievement towards priority sector target. Similarly,

Non-Banking Finance Companies (NBFCs), Microfinance Institutions (MFIs) and other lending agencies also extend credit to FPOs for business activities.

4. **Leveraging Corporate CSR Funds:** CSR Corporates such as Axis Bank Foundation, Reliance Foundation, SD Tata Trust, etc., provide financial support to FPOs out of their CSR Funds for both, undertaking social & economic activities as also for meeting the formative expenses towards skill building and administrative costs. 7.1.5 Seed Money Assistance- Some of the State Governments are providing financial support to FPOs in the form of limited seed money along with infrastructure facilities to catalyse the business operation during initial period.

In order to support the FPOs in terms of strengthening their capital base, SFAC has launched a new **Central Sector Scheme "Equity Grant and Credit Guarantee Fund Scheme for Farmers Producer Companies"** on 1st January, 2014. The scheme has two major components:

- **Equity Grant Scheme:** A grant of up to Rs.10.00 lakh is provided to each registered Farmer Producer Company (which is registered under the special provision of the Companies Act) to match the member equity raised by the institution. This will enhance the equity base of the FPC and enable it to approach financial institutions for raising working capital
- **Credit Guarantee Fund (CGF):** The CGF will offer a cover of 85% to loans extended by banks to Farmers Producer Companies without collateral, up to a maximum of Rs.1.00 crore.

Way forward

Incorporation of suitable provisions in the bye-laws of FPOs may be done to enhance share capital contribution on an ongoing basis and also a portion of profit may be deposited towards share money. Strengthening of capital base is of utmost importance for undertaking business, to be sustainable and also to access matching SFAC equity. State Govt. should come forward to provide initial seed capital to FPOs to enable them in undertaking business activities.

Sensitizing the Registrar of Cooperative Societies (RCS) and Registrar of Companies (RoC) regarding the concept of FPOs and the need for easing the registration procedures will go a long way in reducing the undue delay and hurdles being faced by the FPOs.

There should be continued handholding, mentoring support to POPIs and FPOs at field level and conduct of refresher training programmes to POPIs not only to overcome the field level challenges but also to keep them abreast of the changing trends and practices.

Taking into account the importance of post-harvest technology, aggregation, grading and standardisation of produce, the processing sector may extend suitable backward integration. Involvement of Corporates in developing post-harvest infrastructure like scientific warehouses, cold chains, seed processing, food parks etc. is essential.

Convergence with State Govt. initiatives and credit support from financial institutions will enable the FPOs in availing subsidies and facilities like common infrastructure, procurement and marketing of inputs etc.

FPOs should be placed on a scale like the SHG movement. The SHGs got prominence because of NABARD's continuous support of the concept and coordinating with Government agencies to create an enabling policy environment. SFAC and NABARD are already engaged in promotion of FPOs on a large scale. Both the organisations can take a greater role in taking forward the FPOs to the next level.

Conclusion

Collectivizing farmers into Producer Organizations (POs) has been considered as one of the way to overcome the challenges faced by the small and marginal farmers. This approach is demonstrating the potential to be more successful in breaking farmer's dependency on intermediaries, and enabling them access better markets and input management system. In the last decade, efforts have been made towards creating and strengthening POs and thus strengthening their position in the mainstream value chain/s.

Greater coordination between farmers, processers, retailers and other players is expected in the supply chain. Now the farmers are not producing to the requirements of the market rather they are relying on the markets to absorb whatever they produce. Therefore, the flow of market information may be given top priority among the members of FPOs right from the beginning of their agricultural operations to marketing of their final produce. Use of ICT solutions to be encouraged and it may be a part of market information system to the farmers.

There is now an increasing understanding that production support activities of small farmers must be linked to market demand and that production activities must be looked at within the context of the whole supply chain and the linkages, or business relations, within that chain. There are strong potential benefits of closer links between farmers and markets.

In many ways, POs form a core part of the strategy to lift small and marginal farmers out of poverty and enhance their competitiveness in agricultural markets. The FPO concept is the need of the hour and it is the responsibility of the Promoting Agencies to strive and make it a realizable dream. POPIs being the key players, their effective participation will ensure immense benefits to the Small and Marginal farmers directly. The success and sustainability of FPOs rests largely on increased productivity, market accessibility, integrated value chain approach, post-harvest.

Over 85% of the Farmers in India are small and marginal farmers with land holdings of less than 2 hectares. The continued fragmentation of land holdings leads to disorganization and it is not viable for Indian farmers to adopt the latest technologies. By organization of these farmers into producer companies, economies of scale can be unlocked and the livelihood of farmers can be improved.

Awareness has to be created amongst banks on credit linkage to FPOs so that FPOs get adequate term loans for their infrastructural requirements and working capital loan for their day-to-day operations, management/storage, and access to quality inputs, technologies and adapting to the changing scenario.

References

1. Annual Report 2016-17, NABARD.
2. Farmer Producer's Organisations - Concept and Practices, 2016. NABARD.
3. Financing Needs of FPOs.2016. NABARD.
4. Policy and process guidelines for FPOs. 2013. Ministry of Agriculture, Department of Agriculture and Co-operation, Government of India.
5. Policy Paper – Financing for FPOs by ACCESS Development Services & UNDP.

Chapter 2

Group Approach to Marketing of Farm Commodities: A Heuristic Note

Prakash Kammardi T N

Agricultural Prices Commission, Govt of Karnataka

Interestingly, the concept of group farming or co-operative agriculture has got not only an ideological backing but also strong economic implications as well as political ramifications in a country like India. The Karnataka Agricultural Prices Commission (KAPC), thebody established to provide technical guidance and pragmatic suggestions on aspects related to competitive marketing and remunerative prices for farm produces, strongly supports group approach for marketing of agriculture and horticultural commodities in the state. The KAPC had a collaborative arrangement with Indian Institute of Plantation Management; Bangalore to prepare a comprehensive policy and road map to strengthen group marketing to farm produces in Karnataka. The Price Commission's report submitted to the government have effectively emphasized the importance of strengthening the 'collective bargaining power' of the farmers though group approach so as to garner the major chunk of consumers' rupee paid to their produces. The epistle, by focusing on these issues makes a modest attempt to provide heuristic views on the subject.

Ideological Backdrop

Dr. Asoka Mehta, first Planning Minister of independent India and a socialist scholar constructs his thesis forcibly on co-operative farming not only as a means of emancipation of rural life but also to find the very meaning and purpose of life by engaging in such an associative agriculture. Mehta emphasized on associative life against rugged individualism, on co-operation against competitive struggle among men for the means of living. He goes to the extent of abandonment of unregulated competition. His advocacy on co-operative association was empirically drawn rather than speculatively formulated. He states that "socially healthy and psychologically satisfying life can be developed in 'associative' agricultural community that is neither so small as to restrict the range of occupation nor so large as to exceed man's power of associative work" (Mehta, 1964).

However, Mehta was extremely careful of forcible 'collectivization' of agricultural lands as was done in erstwhile USSR. He clearly sees that the small ownership (of land) is a source of satisfaction and pride. To disrupt the umbilical cord that unites a peasant with his land is to invite social tension. Through appropriate incentives and consensus, he believes that a genuine participative co-operative association in agriculture can be established. And finally, Mehta, by looking at Israel-model of collective farming, visualizes a new entrepreneurial type; called 'institutional entrepreneur'to maximize societal welfare by manipulating the market possibilities on the one hand and by drawing sufficient governmental support on the other, so as to build a strong cooperative agrarian system in a country like India (Kammardi, 2011).

Co-operation in Agriculture

In trying to resolve several basic problems confronted by many developing countries including India, the form of 'institutional arrangement' in agriculture is brought into forefront (Peter Dorned and Don Kenel 1977). The approach of land reforms, introduced with much expectation in this regard, though demolished the evil of feudalism in agriculture, could not take the reform forward to resolve the problems like rural unemployment and poverty effectively in India (Dandekar and Rath 1971). The progress of green revolution too was limited by the inadequate focus on the 'institutional arrangement'. The majority of poor peasantry whose holding-size is very small and uneconomical, and who lack irrigation and adequate resources could not participate in the green revolution individually. So a breakthrough not only to enhance the agricultural production but also to garner the legitimate share in the market led economic growth, requires a strategy for creating an effective 'institutional reforms' in continuation of the 'Land Reforms'. The desired structural change can be achieved through 'cooperativisation of existing individual holdings coupled with group marketing in line with what has been advocated by Asoka Mehta.

From the analysis of a few sporadic attempts on cooperative farming societies since independence, it is evident that the performance of co-operative societies with more or less 'homogeneous' membership; in terms of class, caste, creed, religion etc, was better than those with relatively heterogeneous groups. (Sarkar, 1986). However no serious attempts were made to study the progress of cooperative farming in terms of strengthening the collective bargaining capacity in the market arena, let alone to focus on the scope of enhancement of social cohesiveness and community solidarity within the co-operative organization.

A co-operative farming and marketing system can provide wide spread opportunity for the development of entrepreneurial talents and can have important positive consequences for the future economic development of the country as a whole. Hence, it may become the instrument of developing what Dr. Mehta described as 'social entrepreneurship'. Where there is true and wide spread participation, in such process, many individuals will gain experience and be able to test and develop their managerial capacity. In fact such experience may be of a broader and more significant nature than that developed in a family farm system, in that it provides opportunities for dealing both with resource use and allocation questions as well as with human and social organizational issues.

It will be easier for a government to provide support services of all kinds, credit, inputs, extension, marketing etc - to a small number of large units under cooperation, with centralized management, especially once the government has all the appropriate machinery in place. Moreover large units are frequently capable, either alone or in combination with other units, of mobilizing capital, technology and required labour force for establishing agro-industries as well as supply **and value chains** for farm produces which is next to impossible for an uncoordinated group of small farmers.

Agriculture Marketing in India

In India, farmers' produce is generally disposed off in the village, rural/primary market or secondary agricultural market without bringing in to regulated APMC market premises. Regulated markets are managed by Agricultural Produce Market Committees or APMCs, though in some states they may be given different names such as Agricultural Market Committees (AMCs) in Andhra Pradesh,

Regulated Market Committees (RMCs) in West Bengal and so on. Out of total production in case of the fifteen crops considered for the study only 41 per cent has been brought and sold in Regulated Market (APMC yard) as per KAPC report, 2015. More time for transportation, pre-harvest contract with the local money-lenders and traders, quality constraints are the main reasons for selling produce outside the APMC purview. Out of the total produce brought to APMC yard only 53 per cent has received the price higher than the Minimum Support Price (MSP) declared by the government and the rest sold below MSP. As a result, farmers are not in a position to get better return in agriculture, as evidenced from the KAPC study during 2016 that is briefly summarized below:

1. Area under crops such as maize, cotton and sugarcane whose production is increasing both at the national as well a global level has increased enormously in the last decade resulting in over production and surplus in Karnataka. While the areas under crops important from the food security point of view such paddy, ragi, rabi jowar has been declining drastically over the years.

2. On an average, the cost incurred on labour (human, mechanical and animal put together) accounts for about 38 per cent of total cost of cultivation in the cases of 25 major agriculture and horticultural crops. Hence it may be inferred that agriculture is not only becoming high cost enterprise but also transforming in to labour intensive endeavor in a state like Karnataka.

3. When we consider just paid out cost and cost incurred on family labour, crops such as rain fed paddy, Ragi, Bajra, Black Gram, Sunflower and Coconut have not even ensured positive returns to farmers. Considering overall cost (Cost C_3) which comprises of **"opportunity costs"** such as rental value of owned land and managerial cost, except red gram, fruits and spices, cultivation of all other crops have turned out to be loss making proposition in the State.

4. The MSP declared by the Union Government is less than the actual cost incurred by the farmers in majority of cases and thereby justices have been denied to the farmers in a state like Karnataka. In squeal to above, during the last three years with an intension to curb inflation Union Government has enhanced the MSP by paltry around five percent annually.

A country like India is still lacking considerably in the field of agricultural marketing. These inadequacies are becoming more acute with the significant changes taking place in agri-food systems in domestic and overseas markets. Importantly, the attainment of competitiveness is becoming increasingly dependent on the capacity of the country to develop effective and efficient agricultural marketing. Presently, agricultural marketing system in India suffers from number of constraints which are related to either infrastructure or government regulations or technology or poor market intelligence and finally to low producer's realization. The existing marketing infrastructure is often below required levels both in terms of capacity as well as quality of the facilities.

The Challenge Ahead

Linking small primary producers with markets has been identified as one of the major issues in policy and practice in improving livelihoods for millions of poor in a country like India. National Farmers Commission, under the Chairmanship of Dr. Swaminathan recommended availability of Markets to be within 5 km radius (approx. 80 sq km). Small producers have many competitive advantages like lower cost because of family labour abundance, higher capability in working capability, and traditional knowledge that can be harnessed for many sectors. The only threats they face are: the demand for quality, standardized products in national and global markets. Typically, farmers complain a lack of market for their produce, while processors, exporters or supermarket retailers complain of a lack of adequate supplies of quality produce. This marketing paradox is present because often, buyers do not reach out to explore new suppliers or farmers lack an understanding of markets as well as the ability to identify new markets or to take advantage of such opportunity with value addition activities like grading, cleaning, sorting, packaging and primary processing.

As elsewhere in the developing world, in India, small farmers' livelihoods are being threatened due to the liberalization and privatization of Indian agriculture and the increasing interest of private capital in the agribusiness sector. The withdrawal of the state from productive and economic functions, and changes in the organization of marketing channels, present new challenges for small and marginal farmers. In this environment of greater instability and competition, organization and collective action can help to enhance farmers' competitiveness and increase their advantage in emerging market opportunities.

Globally, and more so, in the developing world, including India, numerous types of market linkage arrangements are available. But, most of these do not ensure that small growers are part of these arrangements. Many market linkage arrangements just provide another alternative to the primary sellers without any commitment to buy or add value as is the case with most fresh F&V retail chains in India which procure only 'A Grade' produce without any contract and the producer is left to sell the rest of the produce in other channels. Most of these channels also deal with individual growers and there have just been only few attempts by private corporate players encouraging the formation of grower groups or associations through the producer company route in India. Hence there is a need for Reform in Agricultural Marketing and empower producers with knowledge, information & capability to undertake market-driven production. Also, it is essential to recognise the importance of small producers in the value chain and facilitate their inclusion in the wider economy

Group Approach to Agricultural Marketing

Indian producers are unable to realize optimal value from their produce and progress further due to fragmentation of land holdings and lack of grass-root level organizations. Also, processors are not in a position to get quality raw material in right quantity. Besides the share of producer in consumer price is abysmally low due to the presence of middlemen. To overcome these problems, direct marketing, contract farming, direct linkage with Retailers/ Processors/ Exporters and market oriented production are some of the approaches. Recently many initiatives have been taken by NABARD and other organizations to promote and involve Self Help Groups, Joint Liability Groups, Farmer clubs, Farmer Federations, SHG Federations, Producer organizations such as Producer Companies, Producer cooperatives, etc in direct marketing of the farmers' produce for better price realization. The more significant govt. initiatives include Horticultural Producers' Coop. Marketing & Processing Society (HOPCOMS - a cooperative) in Karnataka and SAFAL F&V project of National Dairy Development Board (NDDB) in Bangalore.

Collectivization of producers, especially small and marginal farmers, into producer organizations has emerged as one of the most effective pathways to address the many challenges of agriculture but most importantly, improved access to investments, technology and inputs and markets. The Department of Agriculture and Co-operation, Ministry of Agriculture, Government of India has identified FPOs registered under the special provisions of the Companies Act, 1956 as the most appropriate institutional form around which to mobilize farmers and build their capacity to collectively leverage their production and marketing strength. An organization can be called a FPO, if it is formed by a group of primary producers. Producers are primary shareholders in the organization. It deals with business activities related to the primary produce / product / related inputs. It works for the benefit of member producers.

In this background, Karnataka Agricultural Prices Commission has assigned a project to Indian Institute of Plantation Management (IIPM) Bangalore to undertake capacity building for those who are managing the group marketing of agriculture and horticulture commodities currently in Karnataka and based on which to assist the KAPC to bring out a comprehensive policy and road map to strengthen group marketing in Karnataka.

The findings of this collaborative work are summarized below:

1. Organizing farmers into formal or informal groups to meet the volume requirements and strengthen farmers' bargaining power is very crucial. Group Activity is more effective for the benefit of the members of the group than the individual efforts. Informally formed small groups called as self help groups have exhibited their strengths in various fields including agriculture, in improving financial conditions of the members.

2. The following illustrative options could be available to the members of a Producers' Organization (PA): (i) take their produce to a State Agencies Warehouse or to APMC warehouse or sell in APMC. (ii) Obtain finance against their produce through the Warehouse Receipt Financing from banks for the produce store in the State Warehouse or other accredited warehouses. (iii) They can sell the produce on spot or future market depending on price situation known through the warehouses. (iv) The produce can be sold to direct marketing license holder who may be a trader, exporter, processor or retail chain operator. (v) Benefit of effective supply/value chains can be garnered only through PA.

3. Producers organizations (PO) could be one of the best alternative for enabling farmers/producers to get better remuneration for their produce because it enables aggregation of the produce and in turn gives the necessary bargaining power to get better price.
4. To strengthen the Producer Organizations and to make them play an effective role in alternate marketing the following areas need attention: Credit availability, Capacity Building, Alternatives to Equity, Venture Capital Fund, State Support to Producer Companies (PCs) and Convergence of various schemes to PO.
5. Farmers Producers Organizations can help smallholder farmers participate in emerging high-value markets, such as organic marketing, export market and the unfolding modern retail sector in India. The success of FPOs or group marketing, however, depends on farmers' commitment to the group, integrity and quality of the leadership, its acceptance within the community, as well as the market environment. It has to be economically beneficial for the participating farmers to market their excess production through the FPO. At the same time, appropriate knowledge needs to be provided to the FPOs/ Federations to better utilization of excess production from within the community and to maintain linkages to the target markets.
6. NGOs and other supporting agencies could play a crucial role in setting up a successful FPO model which is a lengthy and demanding undertaking, which cannot be done on smallholders' individual initiatives alone.
7. FPOs have the advantage of flexible production methods, they integrate local knowledge, are locally embedded, they are more sustainable with regard to the environment and to the livelihoods of the people involved. In this way, they empower small farmers while giving them the opportunity to deal with contemporary market actors and to enter high-value markets within the Indian economy and abroad.
8. And finally, the collective strength of farmers could enable them to increase their competitiveness through easier access to credit and technology, reducing costs of distribution and providing greater marketing power and negotiation capacity. FPOs could emerge as one of the most effective pathways to address

agricultural challenges. Through adequate policy and infrastructure support these aggregators can become not only the 'connective tissue', linking supply and demand, bridging a major missing link but also become instrumental in faster deployment and acceptance of modern agricultural technologies including mechanization.

The KAPC Proposals

To delve further deep in this regard, the KAPC has advocated several strategies for enhancing collective bargaining power of farmers so as to effectively undertake processing, value addition and marketing of farm produces thereby garner maximum share in consumers' rupee. State Agriculture and Horticulture universities shall establish crop specific FPOs on pilot bases in their jurisdiction with the help of Krishi Vignana Kendras (KVKs). Similarly, State Agriculture and Horticulture Departments, MARKFED, Horticulture Federation, Farmers Organizations, SHGs shall establish FPOs throughout the state in collaboration with NGOs and other organizations.

Group Crop Insurance Scheme may be designed to cover these FPOs. The MARKFED and Horticulture Federation shall be made responsible to intervene in the market operations whenever there is glut in arrivals of agricultural produces. Services of Start-up Entrepreneurs need to be utilized for aggregation of quality farm produces and to establish direct link between Farmers/FPOs and Consumers. By undertaking the above measures, it is hoped that the state will usher a revolution in the marketing of agriculture and horticultural commodities in Karnataka.

Acknowledgement

The author is thankful to Dr. Dhanakumar, Director, IIPM, Bangalore and his colleagues for providing useful inputs some are used in the article.

References

1. Dandekar V.M. and Rath, Nilakantha. 1971. Poverty in India. *Economic and Political Weekly*. VI(2). Pp. 9.
2. Mehta, A. 1964. The Peasant and Socialism. Studies in Socialism, Bharatiya Vidya Bhavan, Bombay.

3. Peter D., and Don K. 1977. Some economic and administrative issues in group farming. *Cooperative and Commune*. (Eds). Peter Borner. The University of Wisconsin Press, U.S.A.

4. Kammarardi Prakash T.N. 2011. Indian Agriculture and Asoka Mitra. *Janata*. 66(41). Nov. 2013.

5. Sarkar, S. 1986. India's Agricultural Development - An Alternative Path.*Economic and Political Weekly*. XXI(19). Pp. 10.

❑❑❑

Chapter 3

Farmers' Organizations in Agricultural Development: Need, Status, Problems and Prospects

Mruthyunjaya, NAIP, ICAR

Need

Agriculture is the largest unorganized sector in India. It is fast transforming from subsistence to market-led activity. The transformation agenda mainly includes value chain approach making agriculture as a competitive business model, not just a production enhancing project. Aggressive and organized participation of farmers (especially small and marginal who dominate Indian agriculture) in this transformation process can substantially improve their livelihoods through diversified nutrition, employment and enhanced incomes. So far, India has mainly relied on the State and public sector led growth and development of the agricultural sector though micro-level operations (farming per se) have been with individual farmers (Sukhpal Singh, 2010). Farmers normally unite when they have common interest, can clearly articulate their need, can organize different services as well as voice their concerns and act as pressure groups to demand services. Market oriented agriculture pursued through best agri-business practice, diversified nutrition (diet) and employment in farm and off-farm opportunities and higher income has the potential to improve the livelihoods of farmers. But farmers do not have any vibrant institutional model to organize and link with and benefit from market. The issue is whether smallholders with tiny holdings and limited resources can individually capitalize economically the opportunities of market oriented agriculture. Apart from tiny holdings and limited resources, the other challenges of commercializing smallholder agriculture are climate change; globalization; more pressure on natural resources; unfavourable external market conditions; poor rural infrastructure; weak institutions; poor access to technologies, inputs, credit and extension;

reduced investments by Governments; limited engagement of the private sector; poor governance, etc.

One of the suggested strategies to address some of the above challenges is to pursue institutional innovations to promote collective action in organizing farmers into farmer organizations (FOs), groups and associations such as Self Help Groups, Co-operatives, contract farming, cluster farming, and Farmer Producer Companies into modern value chains (Chamala and Singhi). Some of these initiatives are led by private sector following the Government's initiative of pro-poor policies such as amending the APMC Act, promoting FPCOs, allowing contract farming, strengthening cooperatives, promoting warehouse receipts, etc. FOs provides backward linkage (input, credit, technology) and forward linkage (production, facilities, market and value chain) and feed- back to research and extension. They are institutions of participatory governance with grass root structures constituted by smallholder, able, motivated sufficiently independent farmers as building blocks, representing their interests and with certain level of accountability to them. They have to work together as a recognized legalized entity in order to strengthen their voice for articulating their needs, for lobbying and taking advantages of economies of scale.

Status

Of the different types of FOs, Cooperatives are historically important in rural India contributing to agricultural development and farmers' welfare. It started with Cooperative Credit Societies Act 1904 under which rural credit societies were formed. Now cooperatives have extended across all over economy covering currently an estimated 230 million members, nation-wide. Of the total cooperatives, nearly 80% deal with agriculture and 605 of them credit. But every sphere of activity is now covered under cooperatives. The area covered are credit, marketing, processing, production (crops, feed), procurement, handlooms, etc. covering major commodities, as milk, fertilizer, sugar, animal feed, ice-cream, handlooms, rubber processing, arecanut processing and marketing, etc. The types of cooperatives include credit (Primary Agricultural Credit Society (PACS), Primary Agricultural Service Society (PASS), agricultural non-credit societies, agricultural Cooperative Marketing Societies, Co-operative Farming Societies. Dairy cooperatives introduced during 1970s were proved to be very successful. Now about 60% of milk produced in the country is sold through organized milk producers cooperatives. Their activity provided considerable supplementary income to over 70% of rural and urban households. Attempts to replicate the model in other spheres however were not much successful. The slow/selected progress of cooperatives is on account of several reasons some of which include: government's interference and overdependence on government; miss-management and manipulation, lack of awareness of cooperative principle, practices by members and functionaries; poor infrastructure; non-conduct of regular elections; restricted coverage, lack of active participation of members in the management of cooperatives, mounting over-dues in cooperative credit institutions, lack of professional quality management, predominance of vested interest, functional weaknesses, lack of human resource development policy, poor communication and public relation strategies, etc. But, cooperative sector has the capacity and potentiality to harness economies of scale, reduce transaction costs, bargaining power, neutralizing the adverse effects of process of globalization, etc. However, they need to reorient their functions according to changing market demands. The failure of the public sector in several cases and also of private sector particularly in the rural areas has increased the hopes of cooperative sector (www.helsinki.fi/iehc2006/papers2/Das72.pdf).

Alternative institutional mechanisms to organize farmers besides cooperatives were continuously being explored. While exploring alternatives, basic principles (voluntary membership; democratic control; equitable economic participation; autonomy and independence; education, training and information; cooperation and concern for the community) and values (solidarity; democracy; self responsibility; equality and equity) of such organizations were kept in view (farmersorganization.blogspot.in/2017/04/farmers-producers-organizations-values.html). Some of the current attempts to organize farmers into community based or commodity-based organizations include (www.slideshare.net/facegopi/farmers-organizationsfarmers-associations-in-india):

a) Agricultural Technology Management Agency (ATMA) of MoA (GoI) into Farmer Interest Groups (FIGs), and Commodity Interest Groups (CIGs)

b) Private sector, NGOs and Panchayati Raj Institutions into FOs, Farmer Associations, Farmers Unions, Federations, Chambers of Agriculture

c) NABARD into SHGs and Farmer Clubs

d) Line Departments in to Cooperatives, Producer Groups, commodity groups under different Boards

e) ICAR/SAUs into FOs

In the above list, the community based FOs include Agricultural Cooperatives, Framer Clubs, FIGs, etc. The commodity based FOs include Grape Growers Association, Jack Growers Association, etc. Some of the services provided by these FOs in general include:

a) Marketing (input supply, output marketing and processing, market information)

b) Financial services (savings, loans, other forms of credit)

c) Technology services (extension and IT, education, training, custom hiring of machinery)

d) Education services (business skills, health)

e) Welfare services

Small Farmers' Agri-business Consortium (SFAC) of MoA, GoI has been mandated in the formation of FPOs under different schemes of GoI. The initiative started in 2011-12. SFAC by March, 2016 has promoted 694 FPOs out of which 410 have already been registered in different states. Some of the states where they are registered in greater number include MP, Maharashtra, Rajasthan, WB, and UP. But some of the major problems faced by them include, registration of members for want of identity proof, accessing the capital from outside, tax on income, poor professional skills, lack of vision and direction from Board of Directors, operational problems like low share value, poor marketing and value addition expertise, etc. (Sukhpal Singh, 2010).

Another institutional innovation is contract farming (CF). The model has facilitated vertical integration by involving smallholders in asset building through procurement, market information, education, credit disbursement, extension services etc. The corporate sector has played an important role in developing an efficient model of vertical integration from farm to fork. Successful contract farming models include examples in Basmati rice, maize, F&Vs led by private sector delivered high dividend to farmers in the form of reduced transaction costs and grater incomes. The CF models were successful when private sector was retailing the produce, adding value through processing, differentiating the product, branding, exporting raw or processed products and retailing or exporting niche products. Enabling policies and institutional arrangements that attract the private sector to promote organized front-end retailing, develop agro-processing and facilitate exports are keys to the success of CF models (Singh, 2012).

Self Help Groups are yet another set of institutions who served as spring board to reach rural poor with micro-credit contributing to prosperity of smallholder farmers particularly among women. Cooperation, consensus among members, knowledge on linkages and transparency in activities significantly influence the collective performance. SHG coverage in India has seen exponential growth with membership around 7.9 million in 1989 to a membership of 30 million in 2013 (Ayyappan, *et al.*, 2017). Promotion of an alternative approach to SHGs in the form of Farmer Interest Groups(FIGs)/ Commodity Interest Groups (CIGs) and linking to institutional agencies for input supply, output marketing and technology transfer has resulted in improvement of livelihood of farmers. FIGs are the social mobilization of farmers having common interest around a particular class of commodities (Ayyappan, *et al.*, 2017). It is felt that there should be comprehensive policy with institutional framework and adequate incentive for promotion of FIGs (Patil, 2014). Under NAIP, in all 3113 SHGs and 7 APCs were established which have enhanced bargaining capacity, better reach to markets and higher realization prices all leading to enhanced income of farmers (NAIP, 2014).

Features/Problems/Conditions of Success of FOs

FOs is formed to replace/serve as alternatives to existing government dominated cooperative laws by a new people centric/ friendly legislation. The legislation came into force in 2003. FPOs and Producer Companies and other types of FOs were established and they are functioning with varied performance since then. A review of their performance by many studies (Chamala and Singhi, Singh, Rwelamira) provides details of their features, problems and conditions of success. The features of an ideal/successful FO to achieve its objectives are:

1. **Organizational personality**: Mission, structures, incentive and reward system, leadership and management style; able, motivated, and sufficiently independent; common interest; strong conviction that only organized action can deliver the output;

certainty of the delivery of the output on time; strong capacity for ensuring long term stability and sustainability, strong monitoring and evaluation of performance with on course corrections.

2. **Resources:** Enough resources (human, financial, infrastructure, facilities, transport) to organize activities well.
3. **Management:** Distribution of the rewards from collective action in equitable way; availability of the rewards within a reasonable time period; rewards commensurate with the costs associated with continued participation; dynamic and flexible norms and operating procedures; both long term and short term objectives, policy, programs and plan of action; enrolment through membership fee; democracy through regular elections; clear and transparent transactions; Strong and committed leadership with ability for setting vision and direction/strategy; professional program and process management;
4. **Partnership and linkage:** Linkages and network with other farmer organizations; capacity to arrange for major investments and a continuous flow of raw material; opportunistic to fill in the space created by reduced involvement of public sector; improved commercial integration among farmers, merchandisers, processors, retailers within agricultural value chain including improvement in risk transfer, technology transfer, product development, logistics arrangements and market information transfer;
5. **Capacity development:** Capacity for periodic training of office bearers and members and entrepreneurship development; mobilizing members around common concerns and realizing objectives; training in soft skills, management and finance to mobilize them in order to create a substantive constituency operating within an agreed legal and operational framework with a sharpened capacity in developing well researched concept and position papers for policy influence to assert a conducive farming business environment;
6. **Best production and business practices**: Improved farm level competitiveness through better cropping systems, proper harvesting and handling using modern technologies, custom hiring and other services and supplies in rural areas; grading,

sorting, agro-processing, etc. as well as through the volume leveraging of procurement, packaging, storage and transport; sorting, agro-processing, etc.

7. **Policy and Governance**: Conducive external environment comprising legal systems, political support including handholding arrangements by the state/central Governments in the initial years of establishment of FOs, a central agency to coordinate all FOs and socio-cultural, economic and ecological conditions in which the FOs operate with continuous collection and analysis of time series-cross section data.

Studies have clearly shown that the rise and fall of FOs are associated with presence or absence of one or many of these features/ problems depending on the types of FOs and the nature of their activities. Some FOs remained active to obtain subsidies of Government but did not actively work and mobilize local resources to help agricultural development. In some elite rural communities captured all services and resources while the poor and women were left out or received less benefit and hence such FOs withered. In nutshell, the purpose of formation of FPOs is to reap the benefits of economies of scale through aggregation of inputs, output, and various processes; reduce the transaction costs; address the capital and credit constraints faced by the small holders and help them in managing risks and uncertainties; and contribute to the overall health, education and wellbeing of members. But the performance so far is mixed and needs lots of further efforts to overcome the shortcomings and harness the strengths. But as RBI has once observed on the failures of cooperatives that "Co-operatives have failed but they must succeed", similarly, FOs are inevitable in the present stage of agricultural progress and they must succeed.

Way Forward

The review of performance of FOs as institutions of hope for accelerated and inclusive agricultural development in relation to intended objectives suggest that greater attention to be given to set right the following (For detailed strategy and action plan, see Ayyappan *et al.*, 2017).

1. Capacity building including supporting farmer member visits to successful FPOs (management, subject matter of business and best production and business practices, safety and quality

management, entrepreneurship development, resource mobilization and use, intensive ICT use(software, market intelligence, weather intelligence,) of all functionaries and members should be made mandatory in the beginning and periodically afterwards through an adequate and dedicated budget provision

2. Diversification of portfolio to include multiple services including serving as agencies of government services and supply of inputs, certification of seed and other inputs and outputs, organic certification, custom services and supplies, allocation of enough space for FOs, RMC Yards, Banks, Government Go-downs, etc.
3. Following best production and agri-business practices
4. Market linkage, inclusive value chain, supply chain management, supermarket and other marketing innovations, retailing, online marketing
5. Collaboration, partnership and alliances with similar organizations, Govt. agencies, private sector, NGOs, international organizations, etc.
6. Policy support to forge linkage with all intermediaries as partners, financial support/soft loan, emphasis on women empowerment, youth engagement, incubators, taxing, good governance, handholding till sometime, single window system for registration, licensing, GST, FSSAI, etc.
7. Coordination, convergence among all types of FOs by a central agency at the state/central level
8. Good governance with progressive HRD policy, incentives and rewards, transparency in transactions, professional management.

References

1. Ayyappan S, Dixit S, Chandre Gowda, MJ, Letha Devi, G Devi MCA, Subhash, S and Dixit, PK. 2017. Policy Paper on Farmer Organizations: Status and Prospectus.
2. Chamala S and P M Shingi. 2006. Establishing and Strengthening FOs (www.helsinki.fi/iehc2006/papers2/Das72.pdf) (accessed on 25th October, 2017)
3. Farmers organizations/Farmers Associations in India. (https// www.slideshare.net/facegopi/farmers-organizationsfarmers-associations-in-india) (Accessed on 20th October, 2017)

4. Farmers Producers Organization: Values, Principles and Benefits. (farmersorganization.blogspot.com/2017/04/farmers-producers-organization-values.html) (Accessed on 4th November, 2017)

5. J. Rwelmira, Strengthening FOs and CSOs.

(www.afdb.org/..../Strengthening-Farmers-Organizations-and-Civil-Society-Organizations) (Accessed on 20th October, 2017)

6. National Agricultural Innovation Project, Final Report. 2014. An Initiative towards Innovative Agriculture, ICAR, New Delhi, India. Pp.78.

7. Sukhpal Singh: Understanding Practice of Contract Farming in India: A Small Producer Perspective.

(www.ncap.res.in/contract-20farming/Resources/4.1%20 SukhpalSingh.pdf) (Accessed on 15th October, 2017)

8. Suresh S. Patil, G. M. Hiremath and G.B. Lokesh. 2014. Economic Sustainability through Farmers Interest Groups and their linkage with institutional Agencies- An Evidence from Karnataka. *Agricultural Economics Research Review*. 27, 2014.Pp. 141-146.

❑❑❑

Chapter 4

The Success Mantra for Producer Organisations: A Creditor's Perspective

Vinod Chandrasekharan

NABFINS, Bengaluru

Significant changes in the agricultural economy are having profound effects on the viability of rural communities. One way for small and midsize farms to remain viable businesses is to increase income to their operation by participating in profitable value-added processing and marketing activities. A popular strategy being used by producers to achieve this goal is to pool their limited resources through collectivisation through Producer Organisations (POs) which can play a greater role by helping producers earn a larger share revenue realised from their produce.

Thus these POs provide a critical link between the producers and consumers. Recognising the imperative of this institutional infrastructure, developmental agencies and Governments are promoting and nurturing these organisations in a big way. They are considered instrumental in increasing the value generated throughout the chain, such as by ensuring that the quality of products is in line with the standard demanded and ensuring an equitable distribution of the incremental value generated to the producers. They also mobilise support from other stakeholders and enable and empower the producers to negotiate for a fair share of the total profit generated.

The success of such collectivisation, especially in the dairy sector, is well documented and understood and needs no repetition here. The point for discussion here is to understand as to why such a success story could be not be replicated in other value chains.

The discussion is primarily based on the experience gained through the organisation I represent, NABARD Financial Services Limited, a Micro Finance Institution and a subsidiary of National Bank for Agriculture and Rural Development (NABARD) which also has extended credit to such POs in different legal forms, be it a functional cooperatives, financial cooperatives, trusts and producer companies in both the farm and non farm sector.

Till financial year 2017, 120 such POs have been funded till 31 March 2017 to the tune of Rs.120 crore to meet their credit requirements be it for creation of infrastructure or for working capital.

Thus this paper would represent a creditors perspective about what a PO ought to be and what are gaps in these institutions as perceived by a creditor.

Some of the specific characteristics of POs which makes them successful, viable, sustainable orgnisations are summarised below

1. Capital "Trust"

The POs are and should be owned by their members and should be member driven. This essentially means that the rationale for the PO and its context of operations are well understood by its members.This leads to commitment to the causes of the PO and the members trustthe organisation and its leadership.

This proves to be critical for the success of these POs, who generally commence their business activities with aggregation and marketing.

In one such instance, the collective bargain approach could eliminate the middlemen and the producer directly sold their produce to the primary processor thus enabling them to see a significant increase in their price realisations. This had a domino effect. The confidence of the producers increased and they are in the process of setting up a primary processing unit, which will enable them to go up the value chain and realise more income.

2. Utility – The members owned it in letter and spirits

The POs are set up to meet the local aspirations based on the local demand. Hence it is important that there is a felt need for such a PO in the first place. There has to be strong business case for a PO. Only when the business of PO is viable and sustainable, the producers would evince interest in the membership of the organisation.

Hence setting up of POs should be a consequence of an informed choice, shared valueswith atleast a significant part of its members understanding the need and role of such POs in their livelihood. Only then, these institutions would be able to withstand various kinds of challenges to be faced by them during their initial stages of growth.

Butin reality, many a times, the promoters of POs, adopt a myopic approach and compromise on the attributes of shared values and vision. POs remain a promoter's dream with their own values which are not internalised by the producers whom or whose values, the PO is purported to represent.

Further, they (promoters), very often, do not come with an exit plan. It is role of the development funders and the promoters to understand that the role of promoter of a PO is only that of facilitator and mentor and their entry should be with definite exit strategy with a reasonable time period after ensuring that the institution is endowed with adequate and appropriate internal capacity.

It must be remembered that being a social-business-organisation, POs are context sensitive. The attempts to put in place an assembly line for mass production of POs would not yield desired results. Such an attempt would leave only fossils of failed POs.

The general incremental approach of "slow start - rapid scale up" used in the developmental interventions will not work in the case of POs. POs should be set up only after examining the underlying factors like the need, absorption capacity of potential members, business case for a PO, social, economical and political context in which the PO would operate as these aspects are very critical n the sustainability and viability of a PO.

3. Balancing Act – Transparent and sound business practices

PO being a social animal, comes with their own set of challenges which are inherent to their membership. It must be borne in mind that this institution is an association of competitors who have come together for the purpose of potential additional benefits which may accrue to them through their membership with the PO.

In such a scenario, it is an arduous task to maintain the balance between the demands of the market for increased efficiency and value and the incremental realisation at decreased cost by the members for their products and services.

It is in this context that the governance of PO becomes very important as it has a direct impact on their operations and sustainability. Member communication, transparent decisions based well informed rules and regulations clearly defining the rationale for internal allocation of costs and benefits, both among the members, and between the members as a group and the PO should be developed through a consensus among members. This is key to the long term success of the POs.

4. Networking & Communication – Shared values and understanding

One great attribute of a successful PO has been its ability to network with all its stakeholders effectively be it Government, Financial Institutions, promoters, business associates or its members. Many a times, it is observed this role is effectively managed by the promoters, which generally is an external agency.

While this approach is good for a nascent organisation, it is observed that this proves to be the nemesis for the POs as they are left high and dry on the exit of the promoters or alternatively they become permanently dependent on the promoters for this.

The other aspect is the deliberate attempt of the promoters to insulate these POs from the market. Such an approach, mainly based on a misplaced notion that the producer by themselves may not understand the nuances of marketing, leads to a situation of dependency.

On a longer term, it is important that undistorted market signals are transmitted to individual membersand the PO as a whole so that they can and are able to able to respond. 'Shielding' members from market signals leads the POsto failure.

In one of the instances of such funding, the entire networking, marketing and sale of the produce of the PO was being handled by another subsidiary of the promoter and the PO in this case is playing the role of a mere contract manufacturer. The only advantage being the increased assured income to the primary producers. Only a small portion of the 'super profit' of the marketing arm was shared with the primary producers. The insulation of the PO from the market makes the sustainability of the PO on a stand-alone basis doubtful. Further this arrangement may see a strain on a longer term, when the primary producers realise the folly of such an approach.

5. Leadership and Managerial Capacity

POs need to have the management and organisational capacity to play an intermediary role between producers and their customers. Whether the members themselves carry out management tasks or whether they hire outside professionals, managerial ability needs to be strengthened.

POs participating in value chains need to have expertise in marketing, in the technical aspects of production, in input procurement and distribution, in meeting statutory and regulatory requirements, and in financial management.

When outside professionals are brought in to manage the organisation, the governance relationship between the management and the board of directors requires extra attention. The members of the board will only be able to direct and control the management if they themselves have sufficient knowledge of marketing strategies and customer requirements.

Hence, Institutional Capacity Building (ICB) should be a part of the business and development plan of the PO so that the leadership of the PO is able to identify the opportunities in the market and initiate timely and appropriate action to meet the demands through the resources which could be generated either internally or from external sources.

In one of the interesting cases, the farmers, who on their own without any external support, understood that that there is a good marketing potential. They therefore set up a collective which planned their production among members, aggregated and sold their produce. This lead to higher realisation of price for their produce. Their understanding of the market, players and its dynamics ensured success and provided them confidence to graduate to a producer company which could usher them more benefits.

On the other hand, one of the producer company set up to aggregate and market milk, which was fully designed and implemented by the promoting agency, was a failure as it could garner enough supply thus failing in its commitments as it could not garner enough member support and faulty design of the project.

Structural Challenges

POs face some structural challenges threatening their survival. These are given in the figure given below.

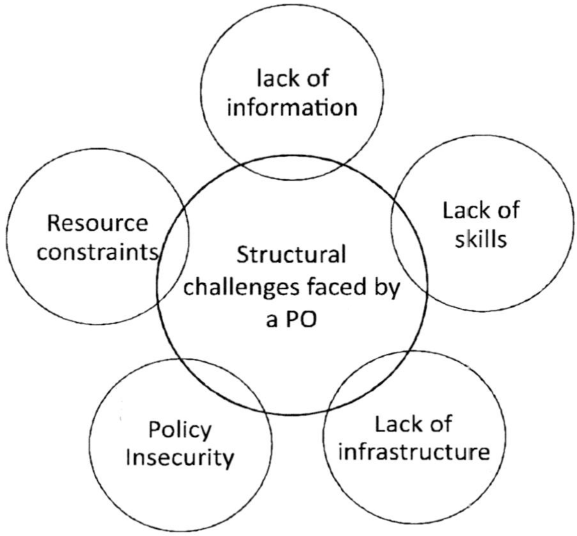

Adapted from: Growing business with small holders by GiZ.

The first and foremost constraint faced by these POs are in raising capital from member producers. Capital from the members are hard to come by, and even if it comes, it does not reach the critical mass.

In case of the Producer companies, there are restrictions on membership to the producer companies, which puts further constraints in raising the capital required. This benevolent piece of legislation attempts to prevent hostile takeovers or hijack by non-relevant players. However, it also has left unintended consequence of preventing much required capital flow to these nascent ventures. This many a times, restricts the activities of POs to aggregation and marketing, leaving them with a lesser share of profits which is ultimate consumer pays to the intermediary. For the time being, The equity support from the promoting agencies / Government may ameliorate this to a greater extent. However a long term solution, which encourages investment in the POs for select non producers with a clear set of guidelines which would protect the POs and its members must be found.

The next major challenge comes in the form of demand for working capital. POs need cash to purchase inputs from dealers or produce from member producers immediately on procurement. In the process the POs are exposed to price volatility; This is addition to other incidentals risk related to the produce like quality, storage, transportation etc.

The way out for this conundrum seem to be is to replicate the success of SHG Bank linkage programme, where the social collateral generated by the group by adopting the principles of "Panchsutra" (regular meetings, regular savings, regular internal lending, regular repayments and transparent books of accounts) is leveraged to extend credit to the groups. The model which is a great success could be replicated with appropriate modification to suit the nature and operations of a PO keeping the underlying principles intact.

These efforts should be complemented by the Government and other development agencies for bridging the gap of information, skill to build the required knowledge through grants or any other financial instruments like soft loan, conditional equity etc. These POs should also be provided to adequate and appropriate infrastructure which could be for the exclusive use of a PO or a set of POs on a pay as you use basis or infrastructure as a service model.

POs should also be provided access to supporting risk mitigation structures like credit guarantees, credit enhancements, insurance, tax breaks etc so that these ventures can establish their roots strongly.

The sheer diversity of produce, producers and their requirements combined with their spatial distribution makes the business of these small producers vulnerable to the vagaries of the market. The adverse economies of scale effectively makes investment in the business risky and many a times unviable.

However, the ascendance of the producers up the value chain is an imperative. It is in this context, the POs become a very critical tool and their success is not only requirement of the times but becomes a necessary precondition for social and economic equity for a better tomorrow.

> India's place in the sun would come from the partnership between wisdom of its rural people and skill of its professionals.
>
> **Dr. Verghese Kurien**

❑❑❑

Chapter 5

The Power of Collective: Farmer Producer Organizations

Singh A K*, Adhiguru P and Letha Devi G**

* ICAR, New Delhi and ** ICAR-NIANP, Bengaluru

Most of the smallholder farmers engaged as food producers in the world is quite staggering, amounting to one and a half billion. Owning farms of a size of two acres or less, they comprise about three-fourth of the world's poorest people, whose food, income and livelihood prospects depend entirely on agriculture. While these smallholder farmers are the primary food producers of the world, they still face huge challenges in the process of attaining commercial success. Given that the agricultural sector too goes through its ups and downs, it is those who can consistently link production with sales who eventually enjoy better prospects. This minority group has been able to take advantage of the better infrastructure and communications, favourable procurement policies of the government, and even the larger number of consumers that is the result of population growth. But then, this is a minority group. The larger picture shows, (especially in Asia and Africa), that 50-70% smallholder farmers are not making the transition from subsistence to commercial farming – they continue to be locked in poverty-stricken conditions and meagre-sized land holdings.

A glimpse into the agricultural history of the world points to a different model of development and modernization in the agricultural sector. It reveals that in developed countries, agricultural modernization and industrial development occurred hand in hand. These had brought in market forces that provided incentives for reallocation of assets, such as land and labour. This created a lean and efficient system of fewer farmers with larger landholdings each. But this is not the case with many of the developing economies where urbanization and industrialization are just not creating enough off-farm jobs that might boost agricultural commercialization. There are also other barriers to a profit-making agricultural model - such as lack of proper local infrastructure, business services that are not robust enough, and lacunae in farmer skills.

This means that overcoming this commercialization barrier requires and upgrading process that includes investment in local infrastructure, strengthening of business services, and improving farmer skills. In many low-income countries, these investments have not been made, and with the reduction in govt. extension services over the past 30 years, most farmers remain unable to access vital technologies and services.

Finding ways to link smallholder farmers to markets is definitely a critical part of any long-term development strategy to reduce poverty and hunger. But research and development work experiences show that this alone is not enough to achieve large-scale poverty reduction and resilience in rural communities. Much effort has also gone into promoting mechanisms to assist the transition from a production-based system to a market based one. These include market analysis, contract farming, certification and strategies to strengthen local business development and support chain investment. Agro-based research has revealed that the effects of market-based approaches have been quite dramatic for those farmers who were linked to high value horticulture.

Variation in small farmer's needs

Research and case studies show that not every smallholder is suited to the same type of market; and that they don't all require the same support services to reach a market. This represents a shift in how extension and advisory services must tailor the type of support they offer to the smallholder. Extension services will play an important role in helping all types of farmers realize their market aspirations. This transition from a production-push to a spectrum of agronomic and business development services is taking place as agencies try to provide services to a full range of smallholder farmer segments.

The agro-enterprise approach

Another approach, which has been employed by multilateral agencies like Catholic Relief Services, is the agro-enterprise approach. This approach is all about shifting from a production-based food-security strategy to a market-oriented approach that emphasizes income generation and profit based on market demand and sale of agricultural products. The term 'agro-enterprise' refers to a business venture, typically small-scale. It may be an on-farm venture or a service that adds value to agricultural goods. An agro-enterprise generally involves groups of farmers and individual actors who provide services within the market chain, and builds relationships with the traders who buy the enterprise's products.

The agro-enterprise approach does not replace the need to work with farmers on improving their agricultural productivity; indeed, the two goals are closely linked. But it does take market forces into account, it does help people grow products that are in demand, and it does help farmers to become more effective so they can reap rewards from their efforts.

Agro-enterprise approaches help farmers focus their skills, knowledge, and resources on producing more efficiently and effectively, thereby increasing their access to food and thus food security. In Uganda, where CRS introduced agro-enterprise methods in refugee camps, displaced persons were able to supplement their irregular food rations by growing cassava to sell at the local market. This approach encourages development agency staff to create market linkages instead of stepping into the market chain themselves. It also empowers farmers to understand their markets and analyse their opportunities. The process empowers the farmers, rather than treat them as passive recipients of new ideas and opportunities.

In many places, microfinance and agriculture teams assisted the self-help group to invest part of their dormant savings into their agro-enterprises so that funds would grow rather than be depleted. Such multi-pronged support serves rural communities better. Integration of such diverse sectors is beneficial for growth of agro-enterprises. They offer prospects for farmers to reinvest in their farms. In the past 20 years, millions of poor smallholders have neglected their farms because they have not earned enough. However, as farm incomes increase, so do the possibilities of re-investment. This can lead to better yields and incomes.

Marketing is a skill, which once learnt, can be applied in a variety of situations. By the time external funding dries up, farmers will have equipped themselves with the skills to meet the dynamics of markets they have already invested in. It will enable them to be more aware of the possibility of switching to other commodities, in accordance with market conditions.

Farmer Producer Organizations

The liberalization and privatization of Indian agriculture saw the State withdrawing from many productive and economic functions, a space that was readily claimed by the private agribusiness sector. The small farmer found himself at the receiving end - his livelihood threatened in an environment of instability, competition and fragmentation of farm holdings. He faced many issues including lack of access to credit and the market, and technology adoption. But rather than a lone farmer struggling with multiple circumstances beyond his control, could he not become part of a collective for mutual support and collective action? Certainly! This is how the concept of Farmer Producer Organization (FPO) was born.

The concept of 'Farmer Producer Organizations, (FPO)' consists of collectivization of Producers especially small and marginal farmers so as to form an effective alliance to collectively address many challenges of agriculture such as improved access to investment, technology, inputs and markets. Department of Agriculture and Cooperation under Ministry of Agriculture, Govt. of India has identified 'Farmer Producer Organizations 'registered under the special provisions of the Companies Act, 1956 as the most appropriate Institutional form around which the mobilization of farmers is to be made for building their capacity to collectively leverage their production and marketing strength. The Department of Agriculture and Cooperation (MoA) has setup 'Small Farmers Agribusiness Consortium' (SFAC), a society under DAC, as designated Agency to act as a single window for Technical support, Training needs, Research and Knowledge management and to create linkages to investment, Technology and Markets. SFAC provides all round support to State Governments FPOs and other entities engaged in promotion and development of FPOs.

The idea behind Farmer Producer Organizations is that farmers, who are the producers of agricultural products, can form groups and register themselves under the Indian Companies Act. To facilitate this process, the Small Farmers' Agribusiness Consortium (SFAC) was mandated by Department of Agriculture and Cooperation, Ministry of Agriculture, Govt. of India, to support the State Governments in the formation of Farmer Producer Organizations (FPOs). The aim is to enhance farmers' competitiveness and increase their advantage in emerging market opportunities. The year 2014 was observed as the

"Year of Farmer Producer Organisations", and slowly but surely, the concept is catching on. The FPO's major operations will include supply of seed, fertilizer and machinery, market linkages, training and networking and financial and technical advice.

Field-based evidence already demonstrates that farmers are reaping the benefits of collective action. For example, in Gulbarga district in north Karnataka, which is known as the 'tur bowl' of Karnataka, the yield was abysmally low owing to use of traditional methods and overuse of inputs. The tur (redgram) farmers never got returns on their investment. Taking note of these problems, a new technology was demonstrated with financial support from SFAC, which helped reduce input cost and increase crop yield. After adopting the new method, one farmer from Gulbarga says that his net increase in income per acre is around 8000 rupees. Another farmer says that he has doubled his income by adopting this new technique, which involved a simple process of growing the seedlings in the nursery, and transplanting them in such a way as to ensure sufficient spacing between saplings and between rows. Farmers have also been able to inter-crop, which earns them additional income. Had these farmers not been part of the local FPO, they would not even have exposure to the new techniques.

A handful of successful examples apart, FPOs need sustenance if it must be mainstreamed in the long run. Help is available through NABARD (National Bank for Agriculture and Rural Development) which is supporting producer organizations, adopting a flexible approach to meet the needs of the producers by setting up the "Producer Organization Development Fund" in April 2011, with an initial corpus of Rupees 50 crore. It provides credit support for financial intervention, market interventions and capacity building in the form of grants, loans or a combination of these.

However, some of the drawbacks that FPOs face is the lack of administrative capacity resulting in poor management of books which leads to issues with accountability and transparency. These factors stand in the way of their accessing finance from banks. To overcome such issues, SourceTrace solutions can help, not only with account management, but also with regard to digitising the traceability / certification process to get premium pricing.

❑❑❑

Chapter 6

Farmers' Organizations: An Overview from KVK Perspective

Chandre Gowda M J and Sreenath Dixit

ICAR-ATARI, Bengaluru

The harsh reality that mere improvement in agricultural productivity and total production are not going to significantly improve the farmers' livelihoods has been realized on account of growing production costs and wafer-thin net returns that the farmers are getting in most of the crops. With the transformation of farming from its subsistence nature to market-oriented production, the opportunities for farmers to reap greater benefits from farming lie in farmers becoming entrepreneurs. Farmer entrepreneurs have to engage themselves in secondary agriculture activities if they were to obtain greater share in consumer rupee. This is not going to be easy considering the fact that, nearly 70 percent of farmers are small and marginal landholders, are unorganized, have diverse resource bases and production practices, have little or no access to capital and most importantly have little exposure to market dynamics. It is much more difficult for the famers to turn into entrepreneurs, especially if they were to remain unorganized and operate as individual farmers.

There have been attempts in the past to overcome this limitation through several initiatives, the most prominent among them being the cooperative movement. The country supported the cooperative movement with several policy initiatives likes Cooperative Credit Societies Acts, 1904 and 1912. The cooperative movement in India owes its origin to agriculture and allied sectors. The farmers generally found the cooperative movement an attractive mechanism for pooling their meagre resources for solving common problems relating to credit, supplies of inputs and marketing of agricultural produce. During 1960s, further efforts were made to consolidate the cooperative societies by their re-organisation. Consequently, the number of primary agricultural cooperative credit societies was reduced from around two lakh to 92,000. There has been a substantial growth in the cooperatives sector in diverse areas of the economy during the past few decades. The number of all types of cooperatives increased from 1.81 lakh in 1950-51 to 4.53 lakh in 1996-97 and the total membership increased from 1.55 crore to 20.45 crore during the same period. Currently there are about 5.45 lakh cooperatives involving 236 million members.

But the cooperative movement has suffered from many limitations. The failure of cooperatives in the country is mainly attributable to: dormant membership and lack of active participation of members in the management of cooperatives. Mounting over-dues in cooperative credit institution, lack of mobilization of internal resources and over-dependence on Government assistance, lack of professional management. Bureaucratic control and interference in the management, political interference and over-polarization have proved harmful to their growth. Predominance of vested interests resulting in non-percolation of benefits to a common member, particularly to the class of persons for whom such cooperatives were basically formed, has also retarded the development of cooperatives. These are the areas which need to be attended to by evolving suitable legislative and policy support.

Several Committees strongly advocated the need to replace the existing government dominated cooperative laws by a new people centric legislation-Choudhary Brahm Prakash Committee (which proposed a model law) (1990), Mirdha Committee (1996), Jagdish Kapoor Committee (2000), Vikhe Patil Committee (2001) and V. S. Vyas Committee (2001 and 2004). The Union Government replaced the existing Multi-State Co-operative Law by a fresh statute – the Multi-State Cooperative Societies (MSCS) Act, 2002. Government of India announced a National Policy on Co-operatives in 2002 - support for promotion and development of cooperatives as autonomous, independent and democratic organisations. In the present competitive scenario, if cooperative enterprises are to serve rural producers, they require an alternative to the present institutional form.

In the parallel developments that took place in rural India, self-help group concept emerged as an informal, village-level institution. The concept entered India in early 80s through the efforts of MYRADA an NGO. SHGs are most popular, especially among women, due to their informal nature. These are unregistered, but, registration is possible and necessary. Social capital – the members' mutual support to achieve collective goals and the savings and thrift activity have eventually helped the groups and group members to launch micro-enterprises as income generating activities. Together, the SHGs are able to empower rural women towards a new horizon. NABARD in consultation with RBI, Commercial Banks (CBs) and NGOs launched the pilot project of linking the SHGs with Commercial Banks. SHG coverage in India has seen exponential growth. With around 500 SHG being linked to banks in 1989, the numbers increased to 2.63 Lakhs in 2001 and 7.9 Million SHG with estimated members of 30 million in 2013.

Since SHG movement was largely successful among women, an alternative approach to collectivize farmers was initiated in the form of farmers interest groups or common interest groups. A Farmer Interest Group (FIG) is a self-managed, independent group of farmers with a shared goal and interest. The members work together to achieve this goal by pooling their existing resources, gaining better access to other resources and to share in the resulting benefits. FIGs are the social mobilisation of farmers having common interest around a particular class of commodities. These groups are also informal entities, but have crucial role in efficient delivery of extension services. The aim is to help farmers groups in increasing productivity, market led production, and quality improvement. The FIG concept has been institutionalized under Extension Reforms Scheme of Government of India under Agricultural Technology Management Agency (ATMA) model.

The Government of India constituted a Committee consisting of experts led by Dr. Y.K. Alagh. On the basis of recommendations of the Committee, a new Part IXA was inserted in the Companies Act, 1956 through "The Companies (Amendment) Act, 2002". The legislation came into force from 6th February, 2003. The Companies Act keeps Producer Companies under the category of private companies but does not impose any restriction on the number of members. The legislation combines the institutional and philosophical strengths of cooperatives (ownership limited to users; limited interest on shares; absence of equity trading, patronage and not capital based) with the flexibility and autonomy of company law. Farmer Producer Company (FPO) offers a new opportunity for the farmers to come together and deal with their problems and harness the opportunities collectively. An FPO or a Producer Company (PC), in short, is relatively new legal entity of the producers of any kind, viz., agricultural produce, forest produce, artisanal products, or any other local produce, where the members are primary producers.

With the increasing emphasis on FPO at the Central and State level, questions are being raised about the viability and sustainability of these institutions in the long run and their ability to promote interests of their members. In the context of 2014 being declared as "Year of Farmer Producer Organizations (FPOs)" by the Ministry of Agriculture, Government of India, greater attention is being given to these emerging bodies. Working through FPOs provides an opportunity to document the benefits of aggregation of farmers into institutions for increased access to income, investments and market opportunities.

KVK perspectives in farmers' organizations

Krishi Vigyan Kendras (KVK) are the district level organizations, working as the frontline extension mechanism of Indian Council of Agricultural Research (ICAR). Each KVK is equipped with essential scientific manpower, farm & laboratories, and educational infrastructure to provide the necessary technical back-up to farmers and development departments. With the financial assistance of Government of India, received through ICAR, KVKs play a crucial role in the agricultural development of the district. KVKs are working with farmers, farmwomen and rural youth in various ways, with focus on agriculture knowledge and resource. Working through farmers' organizations like cooperatives, commodity boards, self-help groups, farmer interest groups, common interest groups and associations has been an integral part of KVK activities. In accordance with the recent emphasis being given to FPOs, KVKs have taken lead in identifying and supporting their pursuit of scientific agri-business development. During 2016-17, KVKs in South India, spread across the states of Karnataka, Kerala, Tamil Nadu, Goa and Puducherry have been working with 94 FPOs. Their status and the role played by the KVKs in these states form the basis of this presentation.

Status of establishment of FPOs

Majority of the KVKs were established in 2016, thanks to the impetus being given by the national and state governments for FPOs. It is clear that the state-level support, was crucial for rapid take-off during 2016. It is worthwhile to mention here the intensive drive initiated by the Karnataka State Department of Horticulture for mobilization of FPOs during 2015 that resulted in establishment of number of FPOs in 2016.

Membership mobilized by the FPOs

Majority of the FPOs have the membership in the range of 200 to 1000 (Fig 1). This indicates that farmers mobilization into FPOs is happening at a reasonable pace and to the optimum extent. This statement is further corroborated by the fact that there are FPOs having more than 1000 members. It is a fact that some of the members have joined out of peer group pressure, but majority of the members are conscious of the need for becoming members of farmers organizations in their own interest.

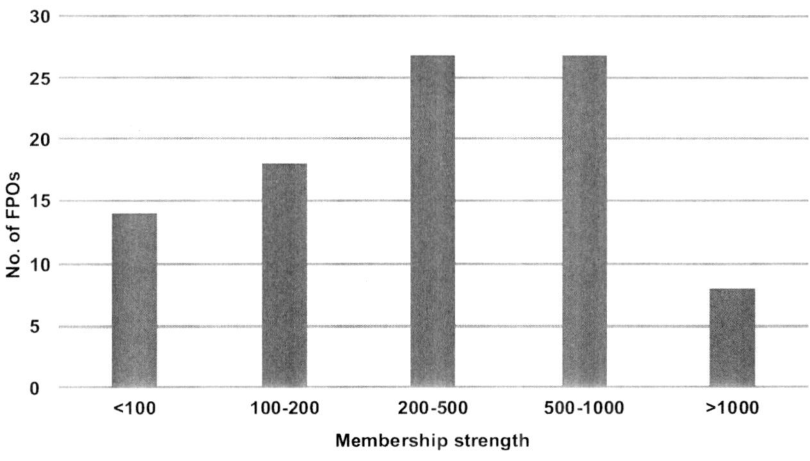

Fig. 1. Membership details of FPOs in South India

Financial resource mobilization

Out of 94 FPOs, 46 FPOs have already succeeded in mobilizing external financial support. Quantum of financial support mobilized appears to be reasonable considering the fact that about ten FPOs could mobilize more than Rs. 25 lakh from the external sources. Altogether, 27 FPOs mobilized more than Rs. 5 lakh from the external sources. Financial health is vital for the FPOs to make business initiatives. Business activities are very crucial to keep the FPO financially viable and to keep the farmers membership intact in the FPO. It is also vital for the retaining the confidence of farmers in the organization.

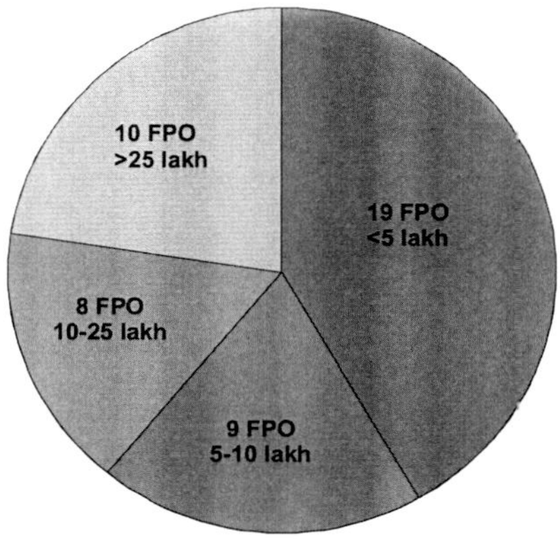

Fig. 2. External resource mobilization status of FPOs in Karnataka, Kerala and Tamil Nadu

KVK roles in FPOs: Participation of the farmers members in the FPO activities depends on the following:

- The degree of farmer's dependence on the organized activity
- Certainty of the benefits
- Equitably of the benefits
- Benefits within a time frame
- Benefits commensurate with costs

Realizing the above, the KVKs are training the management and members of each FPO on the agri-business options available and building their capacity to understand the agri-value chains in these business opportunities. In the process, following roles are being played by the KVKs:

1. ***Technology testing and standardization***: Any new business venture stands on the strength of the technology and innovations. In the event of introduction of agricultural technologies from elsewhere, such technologies require to be tested for their location suitability. This support is provided by the KVKs in the form of On Farm Testing. The cost of such testing and validation is being borne by the KVKs in the interest of technology spread. In certain situations, validation of local innovations is also undertaken through on farm testing and assessment.

Photo 1. A farmer innovation related to scaring of birds and wild animals

2. ***Critical inputs, seeds, planting material, bio-products*:** In the early stages of a technology uptake, farmers face difficulties in fetching the required critical inputs. This initial support is provided by the KVKs in various ways. KVKs do have the required farm and laboratory facilities to take up the production of seeds, planting material, bio-agents, bio-products, equipment, and micro-nutrients for the benefit of farmers organizations.

Photo 2. Farmer participatory onion seed production at KVK Chitradurga

These critical inputs are provided to the farmers as part of demonstrations of technology. Large scale production of these critical inputs are also undertaken as commercial activity, using the revolving fund of KVKs.

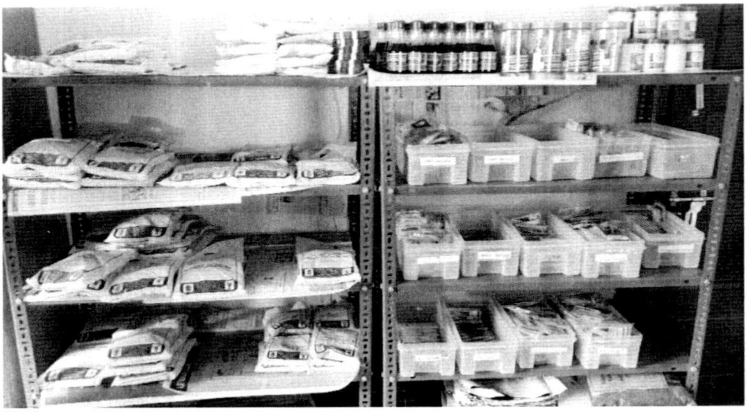

Photo 3. Seeds and bio-products at the sale counter of KVK Malappuram, Kerala

3. ***Technical back-up, capacity development*:** The technology uptake always demand technical competence among the farmers. Such technical capacity development is provided by the KVKs. Such a capacity development is arranged through training and exposure visits to the FPOs.

   It could be in the KVK campus or in the fields of successful farmers, research institutes and development departments.

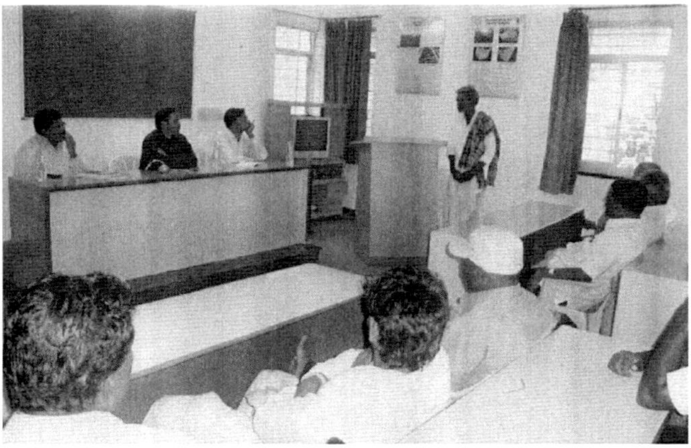

Photo 4. Understanding the technical needs of Interaction with onion seed producers' association at KVK Gadag

4. ***Incubation centres***: Besides the initial capacity development of potential and budding entrepreneurs, KVKs also serve as the incubation centres. The technical handholding during the incubation period will fine-tune the technical competencies and also act as confidence building measure.

   These facilities are provided on the basis of mutually agreed upon terms and conditions.

Photo 5. KVK Mysuru is serving as incubation centre for many enterprises

5. ***Share resources, laboratories, diagnostic facilities***: The farm and laboratory facilities available with the KVKs are not always fully utilized due to limited manpower and the workload that arises out of their engagement in various other activities. This enables the KVKs to share the available resources, laboratories, diagnostic facilities for the benefit of entrepreneurs. Several revenue-sharing models have been worked out by the KVKs to encourage full-utilization of the resources.

Photo 6. The Bio-products lab of KVK Wyanad, Kerala utilized by SHG members

6. ***Producer Organisation Promoting Institute (POPI)*:** The KVKs are also serving as Producer Organisation Promoting Institute (POPI) as per the request of NABARD in selected districts. No doubt, this is the huge responsibility considering the severe constraints related to man power availability in the KVKs. However, in the interest of farmers and agri-business development in the district, KVKs have taken up this responsibility in exceptional cases. This is particularly so in the districts where institutions capable of taking up this responsibility are not yet fully developed. Financial support provided under PRODUCE fund of NABARD is used for this purpose.

Photo 7. KVK Karur, Tamil Nadu is the POPI for few FPOs in the District

7. ***Constraint analysis and assist in policy making*:** KVKs are playing pivotal role in facilitating state specific policies related to promotion of farmers' organisations. During 2015 state level workshops were organised in Kerala (at KVK Thrissur), Karnataka (at KVK Gadag) and Tamil Nadu (at KVK Namakkal). The Agriculture Price Commission, Govt. of Karnataka and NABARD Regional Office in Karnataka used the feedback and suggestions of KVKs in fine tuning the strategy related to FPO development in Karnataka. State level and university level interactions were arranged to facilitate state specific policies. The State Department of Horticulture, Govt. of Karnataka and ICAR are in the process of signing MOU for enabling the KVKs to support the Farmer Producer Organizations with necessary technical back up for capacity development of FPOs.

Photo 8. State-level policy facilitation in Kerala, organized at KVK Thrissur

Conclusion

Krishi Vigyan Kendras are the district level agricultural science centres having several unique features. A team of multi-disciplinary scientists and the state of the art farm and laboratories are the two important unique features on which a credible image for KVKs as agricultural knowledge and resource centre has been built. The KVKs are serving the technical needs of farmers and other stake holders keeping the agricultural science at the forefront. Ture to their image, KVKs are supporting the Farmer Producer Organizations in the states of Karnataka, Kerala and Tamil Nadu in multiple ways. These roles range from being promoting institutions to policy facilitation. However, most important is the technical backup role that can propel the FPOs towards planning, executing and attaining a viable agri-business specific to each organization. This role is not a one-time measure but a continuous process, helping the FPOs and their agri-business plans evolve continuously.

References

1. Anonymous. 2007. Workshop on Producer Companies, Pradan NGO, 2007. http://www.pradan.net index.php? option=com_ content &task=view&id=122&Itemid=1 Accessed on 15 Oct 2016.

2. Fernandes V. 2015. Agriculture's big hope: Farmer producer companies learn to tame middlemen.http://www.firstpost.com/business/agricultures-big-hope-farmer-producer-companies-learn-to-tame-middlemen-2417802.html accessed on 05 Oct 2016.

3. GOI. 2013. Policy & Process Guidelines for Farmer Producer Organisations, Government of India. File:///C:/Users/admin/ Desktop Final % 20 Edited% 2007. 07.2016 /Reviews/FPO-Policy & Process-GuidelinesDAC2013.pdf

4. Mondal A. 2010. 'Farmers' Producer Company (FPC) Concept, Practices and Learning: A Case from Action for Social Advancement.'*Financing Agriculture*.42(7): Pp.29-33.

5. Nalini B., Turaga R. M. R. and Vaibhav B. 2016. Farmer Producer Organizations as Farmer Collectives: A Case Study from India. http://www.iimahd.ernet.in/assets/snippets/workingpaperpdf/ 10539809132015-01-05.pdf accessed on 15 Oct 2016

6. Sharma P. 2010. 'Promoting Farmer Producer Organizations to Mitigate Risk and Improve Market Access: Lessons and Challenges', *Financing Agriculture*. 42(7): 22-25. http://manipursfac.com/wp-content/uploads/2011/09/Article_MD1.pdf accessed on 19 Oct 2016.

❑❑❑

Chapter 7

Farmers' Organisation: A Sustainable Starategy for the Development of Farmers in India

Narayana Gowda K and Narayanaswamy K C

University of Agricultural Sciences, GKVK, Bengaluru

Agriculture is becoming more challenging today in the history of Indian Independence. India feels proud of achieving the record food production of 273 million tons during 2016, despite serious drought in many pockets of the country, more so in Karnataka. The productivity and production of fruits and vegetables, milk and other farm based produces have increased several folds. In many cases, the surplus production has taken place particularly among perishable commodities because of lack of database information. These achievements are seen despite many serious problems the farmers are facing today.

The most important problem is not finding a remunerative and assured price at the easy reach for all the farm produces despite the country is nearing 70 years of independence except in case of milk. Although the profit margin in case of milk, the single largest enterprise is very narrow, still farmers are continuing with the enterprise because of procurement of milk at the village level timely as well as transparency in payment. There are other associated problems namely; divisions and fragmentation of holdings, indiscriminate increase in cost of production, non-availability of labour, timely non-availability as well as at the easy access, quality critical inputs, inadequate and timely technical support, credit, insurance, decline in natural resources particularly underground water and soil fertility. Yet other serious problems are: increasing nucleus families and inturn insecurity, increasing constant migration of youth to urban areas, not getting alliance to prospective groom, increasing social problems, poor infrastructure facilities for education and health which are impacting on declined interest in rural life.

There is an absolute need for rebuilding the confidence of rural people in general and farmers in particular. This is possible only when the farming becomes profitable enterprise. Farming becomes profitable only when the farmers are organized, based on enterprise or combination of enterprises. The necessity of organized group becomes essential in view of large number of small and marginal farmers as well as to undertake secondary agriculture which is a difficult proposition for the individual farmer. Milk Producers Co-operative Society (MPCS) is one such organized group successfully functioning in the country. Organised group not only addresses ensuring profitable price for the farmers but also employment to many unemployed and underemployed people in the rural areas besides range of other benefits.

Types of organised groups

There are different types organized groups functioning in India, most majority of them in recent years. The most commonly observed ones are: Self Help Groups, Commodity Based Associations, Farmer Interest Groups, Farmer Producer Organizations, Farmer Producer Companies and so on. Although the nomenclatures used are different, but the centre of focus is bringing together all willing people to work together to address their individual as well as common problems. These initiatives will bring in harmony and rebuild confidence in rural life which is the need of the hour. What Mahatma Gandhi observed "that India lives in villages, if villages are developed, India will develop". Dr. A.P.J. Abdul Kalam, Former President of India said 'PURA' (Providing Urban Amenities in Rural Areas) is possible only when the rural income improves as well as investment on rural infrastructure is increased substantially.

Keeping aforesaid issues in view, there is absolute need for organized group for the sustainable development of rural areas in general and farmers in particular. Even to ask for legitimate demand what the rural people want, there is need for organized group.

What are organized groups

These are the grass-root level autonomous registered institutions established around single enterprise or group of related enterprises to address end to end issues in order to improve production efficiency, minimize cost of production, enable value addition and processing and enhance income of its members. In the process, it is expected that the farmers and the group to become independent in managing an enterprise profitably. These organized groups can address nearly 10 of the backward and forward linkages.

A. Benefits of Organized groups

- Registered bodies under the Registrar of Firms & Societies
- Autonomy, flexibility and transparency in the system
- Strengthens backward and forward linkages
- Promotes division of labour , specialization and shared labour concept
- Resource sharing including machinery and infrastructure
- Provide equal opportunities for all sections of rural society for knowledge sharing, input and marketing management.

Backward linkages

1. **Technical support:** Technologies are changing everyday particularly in hi-tech, even among low cost interventions they need to be updated regularly. In the yester years, Gramasevak was in contact with every farmer and use to visit regularly every village. In Karnataka, the head quarter for the Gramasevak (later on re-designated as Agricultural Assistant) was at panchayat level. After the reorganization, the head quarter was shifted to Hobli level, which is designated as 'Raita Samparka Kendra' (RSK). Generally, there will be four to five panhayats for each Hobli, with two to three Agricultural Assistants under the present set up, it would be difficult to reach every family at frequent intervals. Earlier it was easy for majority of farmers to visit panchayat office which was in the easy reach than going to Hobli level. Further there is constant decline in the manpower working at the grassroots level across the country and in Karnataka, it is less than 40 percent. With the formation of group, the group representatives who are more knowledgeable can visit RSK's and access the credible source and share regularly with the fellow members. Such members could be deputed to KVKs and other concerned institutes for higher training and they can share such knowledge and skill instead of every farmer going to RSK's and

KVK's. Even for database generation these organized bodies will be very useful and it is cost effective.

2. **Critical inputs:** An ensuring quality critical input, such as seeds, seedlings, bio-fertilizers, bio-control agents, agricultural implements, etc., at an easy reach of an average farmer is bigger problem. Instead of each individual farmer running around these agencies/ places, one or two office bearers can take up ensuring them timely besides minimizing overhead costs with required quality.

3. **Availing subsidy and other facilities from various Institutions:** Knowledge of various schemes andavailing subsidy from different institutions is a challenging task. Many a times, a farmer has to run around several people both officials and non-officials to avail such a benefit. Many a times, more than the benefit, the expenditure incurred and time spent will be several folds. Inadequate staff is another reason for ensuring each of these benefits to the beneficiary. Two to three office bearers who are more knowledgeable with the procedures can attend these activities timely.

4. **Credit and other support from financial institutions:** Even highly educated persons find it difficult to avail loan and other facilities from financial institutions. The nature of application, the number of records and filling the application form with required information is a difficult job even for an educated farmer. Many farmers avoid taking loan from the Commercial Banks for these reasons. The representatives of organized group can shoulder these responsibilities of getting loan as well as ensuring timely repayment.

5. **Insurance:** It is more being said than availed because of complexity involved both in submitting the filled in application form timely as well as availing the benefit. The representatives who have more knowledge in this endeavor can do these responsibilities for the benefit of fellow members.

6. **Use of farm machinery on custom hire basis:** In view of decline in the availability of labour and drought animals, the dependency on farm machinery has increased in the recent years. Individually, it is uneconomical to own, repair, maintain as well as use them to fullest potential. This arrangement would ensure

timely farm operations, increased productivity and minimize cost of production.

7. **Shared labour:** The shared labour arrangement was in vogue in the yester years, but hardly seen in the recent years. With the increased nuclear families the necessity of this concept has become all the more important now. Farmers feel comfortable to work with others when there is a common enterprise. This will enable undertaking skill based activities more precisely. The concept of helping each other increases and minimizes social problems which are major threats in the recent years.

Forward linkages

1. **Premarket arrangements:** Permanent arrangements can be made for common threshing, grading, packing, storage and transportation with the involvement of members who are skillful in these activities.
2. **Marketing:** The biggest challenge for the farmers is marketing. Individually they find it difficult to address the problem of marketing. Choice of marketing place could be decided jointly based on market intelligence and past experiences. Facilities available with various institutions could be availed by group discussions. Two or three experienced members can take up the responsibility with common transport and acceptable guidelines.
3. **Value addition and processing (VAP):** Value addition and processing cannot be taken up individually since it requires important prerequisites such as registration, license, brand name, sufficient capital, space, machinery, specialized training, management skills to manage men, money and market, besides public relations, ensuing quality produce etc. Now Government has brought much leverage to encourage entrepreneurs to take up secondary agriculture by women. Women entrepreneurs may be identified to take up these vocations by availing various support systems available from the government and other organizations.

Effectively addressing forward and backward linkages by the group and its representatives will directly influence profitable income generation for each producer particularly for small and margined farmers as well as nucleus families. Size of the group may vary from enterprise to enterprise, geographical spread of farm families undertaking such enterprise, low value or high value enterprises, infrastructures required for establishing enterprise and so on. The Rational and Scientific justification has to be given for deciding the size of the organized group.

Rural Bio-Resource Complex (RBRC) Project

The RBRC pilot project funded by DBT-GOI was implemented by University of Agricultural Sciences, Bangalore in Tubagere Hobli of Doddaballapur taluk in Bangalore Rural District in Karnataka State from April 2005 to March 2010 has attempted to address problems of farmers enlisted above in a holistic manner. The project attempted to address end-to-end issues to build robust agriculture based rural economy covering 8,340 farming families as well as non-farming families spread over in 75 villages. The project has formulated and executed strategy consisting of six factors to enhance income and standard of living farmers, which include;

Strategies

- Promotion of capsule of integrated technologies.
- Ensuring effective information support system.
- Providing critical inputs and custom hire services of agricultural machineries.
- Effective functional linkage with research, financial and marketing institutions.
- Market empowerment.
- Establishment of producers/commodity based associations (CBAs).

Associations

The first five factors were formulated in the initial phase of project, while the necessity of last one was identified during the second year of the project. Therefore, it was thought essential to promote commodity based producers' associations to provide enhanced income and reduce overhead costs. Accordingly, 10 producers' associations were promoted during 2006 to 2008 and they are:

1. Rural Biofuel Growers Association, Hadonahalli
2. Jack Growers Association, Hadonahalli
3. Federation of Women SHGs, Tubagere
4. Fish Farmers Association, Tubagere
5. Chawki Rearing Centre, Gangasandra
6. Organic Farming Farmers Association, Karnala
7. Corn Growers Association, Hadonahalli
8. Fruits & Vegetables Growers Association, Hadonahalli
9. Agro Processing Centre, Melekote
10. Flower Growers Association, Hadonahalli

The best-performed ones are presented 1 to 9 and failure was noticed in case of Flower Growers Association. Later on the farmers have voluntarily started 'Sheep Producers Association' and they are considering the start of other such associations.

Interestingly the project has demonstrated 11 percent growth rate in Agriculture besides threefold increase in income. The major contribution has come from Commodity Based Associations.

These associations were established in strategic places of the project area. The most innovative aspect of the model is the use of the underutilized facility with Milk Producers Co-operative Societies (MPCSs) in each village as the base for the procurement of the village produces namely; bio-fuel seeds, fruits and vegetables, vermicompost manure and a range of other produces as well as storage of critical inputs for the timely distribution at the easy access of stakeholders. The planning and implementation of two CBAs are presented in this paper.

A. Rural Biofuel Growers Association (RBGA)

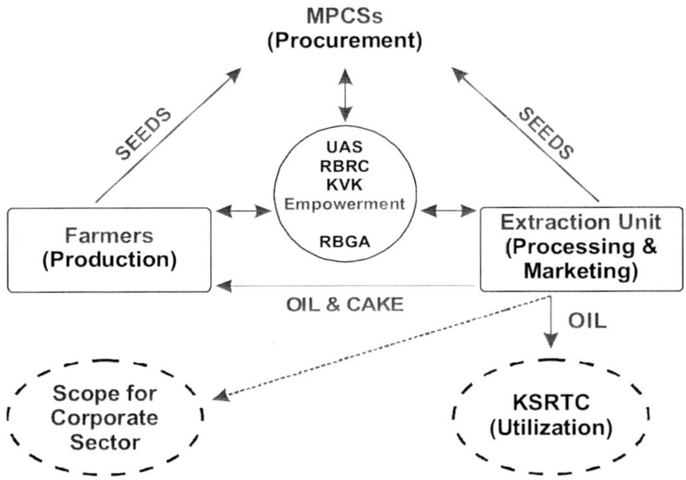

RURAL BIOFUEL GROWERS ASSOCIATION MODEL

The RBGA was established during 2008 after 42 rounds of meetings and exposure visits involving biofuel growers and at present there are 5000 members. A novel initiative is the procurement mechanisms established through MPCS at a cost of 5% overhead cost and provided increased prices from Rs. 5 to 15 /kg during 2008 and presently Rs. 30/kg. The average harvest was 60% before 2008 and it is 100% after the respectable price was ensured. The required seedling were ensured at the village level by training the farm youth to raise biofuel seedlings, later on these farm youth have gone for the production of Jack and Mango grafts. The unit established is providing employment to five of unemployed and under employed youth in the area besides additional employment opportunity for the people in the area. The model addresses end to end solutions for the biofuel enterprise in the area.

Benefits of RBGA

- Availability of biofuel seedlings locally
- Increased area under biofuel plantation
- Better use of non-arable land
- Appropriate technology for dry land

- Improves ecology and environment
- Assured price
- Transparency in weighment & timely payment
- Additional income and employment
- Reduction of overhead charges
- Availability of quality oil and cake
- Cent per cent harvest of biofuel seeds

Various Promotional Activities Undertaken by RBGA

Jack Growers Association (JGA)

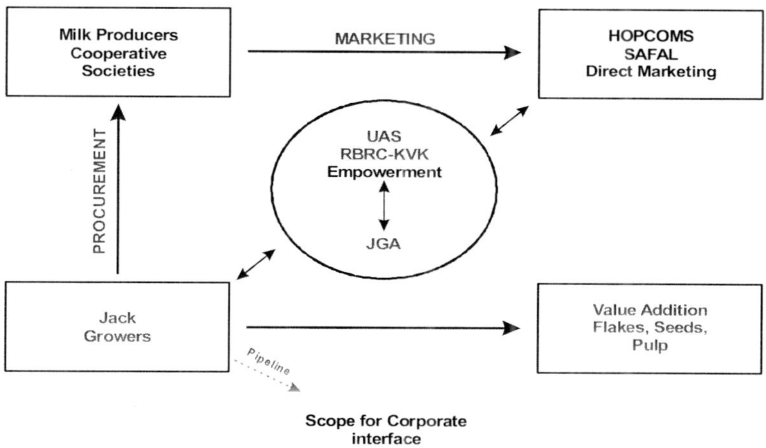

MODEL OF JACK GROWERS ASSOCIATION

Considering the importance of Jack fruit in the area in terms of assured income being a hardy crop with least expenditure on maintenance, huge demand for fruits in the market and the extent of exploitation by the middlemen, the project staff debated on the start of JGA to ensure respectable market and quality grafts for the growers. The farmers' share from consumer was 18 per cent during 2007 and it is 68% at present with an increase of fourfold increasing income besides a range of other benefits. The model is serving as a centre of education addressing end to end issues and efficiently functioning even today with a greater impact across the jack growing belt in the country.

Benefits of JGA

- Large scale production and distribution of jack seedlings
- Scientific harvesting - improved health of the trees and increased consumer preference
- Registration of elite jack variety with NBPGR
- Assured price - Fruit earlier sold at Rs.25 now sold at Rs.100
- Transparency in weighment and timely payment
- Reduction in overhead costs
- Increased income and employment generation
- Value added jack products

Reasons for success and failure of farmers' associations

A. Contributing Factors for Success

- Need for a Farmers' Association should arise from farmers' interest.
- Start of such an organization should be scientifically feasible.
- Nodal department /organization /Board should be responsible for promoting such organization.
- Provide startup funds based on necessity.
- Identify local leaders who are capable of taking the responsibility and their track.
- Provide training to office bearers for managing such organizations effectively.
- Arrange to provide/ensure all the subsidy support system required for the functioning of the organization.
- Enlist all the linkages the concerned for sustainable functioning of the organization.
- Ensure checks and measures for the sustainable functioning.
- Research feedback for refinement and strengthening.

B. Factors Responsible for Failure

- Scientifically unfeasible enterprise.
- Lack of support from concerned Department
- Lack of provision to provide startup funds.
- Lack of good leadership.
- Lack of interest from the local people.
- Lack of budget and other support for establishing the functional unit.
- Weak linkages and poorly defined responsibilities.

C. Spread

Litchi Growers Associations have been established in Bihar on the lines of Commodity Based Associations. Jack Growers Associations have also been established in some parts of Kerala and western ghat region of Karnataka. Two to three CBAs have been established under each KVK for specific crop(s) under the 1Government of Karnataka sponsored IFSD project with a budget outlay of Rs.75 crores in all 29 KVKs of Karnataka from 2011-12 to 2013-14.

D. Breakthrough / Path breaking new initiatives

- Ten Commodity Based Associations established under RBRC project which are being replicated in selected pockets of the country. This initiative has been debated in many developing countries' forums including Round Table Conference held at Beijing in China during 2011 for its large scale implementation.
- The innovative model developed for procuring Biofuel seeds, extraction and sale was very successful. The Asia-Pacific Association of Agricultural Research Institutions (APARI), Bangkok (Thailand) brought out a booklet narrating the success story of Biofuel Growers Market Network. Further, the successful experiences lead to the establishment of Biofuel Park at Hassan, starting of Taskforce on Biofuel and later 'Biofuel Development Board', in Karnataka with a massive budget allocation of Rs. 75 crores during 2011-12.
- The Jack Growers Association gave platform for organizing national and international seminars besides national project on

Value chain on jack fruit with an outlay of 4.65 crores across five major jack growing states in India. The project has come out with identifying best Jackfruit germ plasm, grafting technology which bears fruit in three years, standardized 1\value added products which can be taken to global market and so on.

Conclusion

The sustainable agricultural development rests with mobilizing local people to take up the responsibility of adopting appropriate technologies for realizing potential yield and best price for their produce as well as undertaking value addition and processing which are effectively addressed through the start of need based commodity based associations. The necessity has become all the more important now than ever before since many of the public extension systems are drying up, small and marginal farmers as well as nucleus families are increasing and the confidence level of these families are decreasing, secondary agriculture becomes increasingly essential to provide employment opportunity and added income to rural people in general and farm youth in particular.

ARYA, which is implemented across the country, the major emphasis, given, is group-based enterprises, which will be an important momentum in the coming years. The establishment of effective functional linkage between organized groups with KVKs and other related institutions has strong influence on sustainable development of farmers in India.

Acknowledgement

1. We profusely thank the University for having given an opportunity to work for RBRC project.
2. Special thanks to DBT-GOI for having given this project to UAS (B).
3. Remembrance to Scientists, RAs, SRFs and office staff who have worked for the project.
4. Thanks to all the collaborating institutions actively involved in the project implementation.
5. Regard to all the Local leaders, Farmers, CBAs and others who actively participated in the implementation of the program.

❑❑❑

Chapter 8

Farmer Producer Organizations: Promising Institutional Vehicles for Economic Development of Small Holder Farmers and Initiatives of Governement of Karnataka

Maheswar D L and Ashok S Alur

University of Horticultural Sciences, Bagalkot

Agriculture continues to be the major source of income, livelihood, employment and a predominant business option for the people in the rural areas in India as more than 50 per cent of people continue to depend on agriculture. Smallholder agriculture is facing several challenges and constraints and there is a need for developing highly radical and innovative policies and mission mode approaches to address these issues. This mainly requires concerted efforts of integrating farmers that have fragmented holdings and retain youngsters in the rural areas who are migrating to urban areas. This grave area of concern can be addressed through innovative institutional models and approaches. India is the second largest country to have highest number of small holdings after China which indicates that India has a greatest challenge of making these small and marginal holdings truly productive and assist them serve as the main source of income and livelihood for millions of small holder farmers dependent on them. Historically various institutional models have been prevailing in India and various new models area are also being tried in to integrate small farmers with the value chain. Among the prevailing models, the most common model is the producer's cooperative. Though this model has been able to bring farmers together for getting organized themselves as a cooperative but the level of cooperative experience in Indian context has not been a very successful except few cases. The government departments have largely promoted the cooperatives in agriculture with a focus on welfare of the members rather than the focus on doing the business on commercial lines and scales. Realizing the need for giving the business and commercial outline to the cooperatives, the Government of India enacted the Producer companies Act by incorporating a new section IXA in the Indian Companies Act.1956 through an amendment to the Indian Companies Act-1956. This enables the registration of the producer companies with the Registrar of Companies. There is also a provision to enable incorporation of cooperatives as companies and conversion of existing cooperatives into companies, while ensuring that the unique elements of the cooperative business remain intact in the new legislation.

Farmer Producer Organizations

The collective of farmers in the form of a FPO is gaining popularity due to several advantages it carries in comparison to the conventional model of farmers' cooperatives. The Producer Companies Act enshrines the ethos and basic tenets of cooperatives and infuses a greater professional approach and attitude into the management of FPOs. Usually the member farmers form the FPOs with the equity contribution. The act provides a provision for the FPO with appropriate framework for owning the company collectively by the farmers themselves. The day to day operation of the FPO is expected to be managed by a group of professionals, hired from outside, under the direction of the Board of Directors elected or selected by the General body of the FPO for a specific tenure. Since farmers are the equity holders of the FPO, it provides an appropriate framework for owning the company by the farmers themselves.

Characteristic features of FPO

FPOis a corporate body registered under the Indian Companies (Amendment) Act 2002 whose ownership and membership are held only by 'primary producers' or 'Producer Institutions' and the equity cannot be traded but may be transferred to other members. The transfer of equity can be done only with the approval of the Board of Directors. The liabilities of FPO are limited to the value of the share capital it has issued and the liability of the members is limited to the value of share capital held by them. The minimum number of producers required to form a FPO is 10 and there is no limit for maximum number of members and it can be increased as per feasibility and need. There cannot be any government or private equity stake in the FPO, which implies that FPO cannot become a public or deemed public limited company and the area of operation for a FPO is the entire country.

Registration of FPO

The FPOs can be registered under different acts or laws prevailing in different states of India. The farmer producer organizations can be registered as cooperative societies under the provisions of cooperative societies act, they can also be registered as societies under societies act, as trusts under the provisions of Trusts Act and as companies under the provisions of companies act. The process of incorporation of a FPO is an important step and has to be initiated by a person or by an organization having enough experience in organizing farmer's institutions. A person or an organization involved in initiating the process of formation of a FPO is called as an initiator or a facilitator who leads the process of incorporation of the FPO and guides the group of farmers through the entire process.

Process of incorporation of FPO

The process of incorporation of an FPO need to be planned properly and executed meticulously. This should also follow the social process as done in case of a community based organizations (CBO). It is also essential to know that a FPO is also a CBO. The FPO also has a shared objective, mutually agreed plan of actions, shared responsibilities and benefits and a mechanism of functioning where the decisions are taken by the opinions of majority of the members. Hence, the processes of building an organization cannot be different in case of FPO than what is generally followed for any CBO. The processes start with the conceptualization of the idea by the initiator about the objective and structure of the CBO that is intended to be formed in a given situation.

Steps in the formation of FPO

The steps involved in formation of FPO are, (a) Setting the goal and objective (b) Developing blue print of FPO (c) Identification of area of operation (d) Finalizing the number of producers (e) Building rapport with the potential members (f) Exposure visit/study visit for the proposed members (g) Assessment of training needs of the proposed members (h) Training of the proposed members (i) Fixing the share value and share collection (j) Membership mobilization (k) Developing the outline for business plan (l) Building Memorandum of Articles of the Association (m) First meeting of the shareholders (n) Selection or election of the members (o) Registration of the company and convening the first General Body meeting.

Steps in Registration of FPO under company's act

The process of registration of FPO can be undertaken after the completion of above steps. The steps involved in the registration of a 'FPO' include (a) Obtaining Digital Signature Certificate (b) Obtaining Director Identification Number (c) Naming the FPO(d) Preparing memorandum and articles of association (e) Registration of company and getting of Certificate of Incorporation.

Business Planning for FPO

Successful operation of FPO is one of the important aspects, which requires the preparation and finalization of a business plan in consultation with the members and experts. While developing the business plan, the FPO has to understand the physical, social and cultural aspects of the proposed area of operation, its potential, requirement of the shareholders relating to farming (supply of quality inputs such as fertilizer, seeds, pesticides, tools, implements, machineries, credit, technical services), surrounding markets and competitors. Demand and supply analysis of products to be sold and product to be required by shareholders would be helpful. A business plan would not only convey the organizational structure, business goals and the strategies to meet them, but also will allow FPO to assess the potential problems and the ways to solve them. Business plan will also help to assess the capital required for the planned business, which, further be required to be submitted to any financial agency (nationalized/cooperative banks) to apply for loans. The first step in business planning is to identify the business opportunities. The agriculture or horticulture based FPOs will basically focus on horti-business or agri-business for the farmers. Identification of specific business opportunity is largely a reactive process. Identification of creative business opportunity can be done through different ways such as brainstorming followed by evaluation and short-listing of ideas.

The business planning for a FPO is very important as the FPO has to operate on a value chain model. Value chain comprises all the activities at different yet interlinked stages that add value to a particular product through the different phases of production, including procurement of raw materials and other inputs. Usually, there are many actors along the value chain for producing, processing and bringing goods and services to end-consumers through a series of sequential activities. The FPO may choose to undertake any one or more than one activity of the value chain. Such activity will be a value chain intervention. The key is to choose the intervention where the value-addition is high, and which in view the FPO keeping in view competition in the market can effectively manage.

The business planning process starts with generation of innovative but practically feasible business ideas. This process has to be followed by opportunities and threats analysis leading to Identification of suitable business opportunities. After identifying the business opportunity a strong marketing plan has to be prepared. The final part of the process deals with the finalization of the financial plan. The business plan provides broad parameters for achieving the goals of FPO. A typical business plan will have Business Description; Industry/Sector analysis; Marketing plan; Operations plan and financial plan.

Financial Management

The financial management of FPO is an important aspect that decides the success or failure of the company. The share capital or equity in the FPO means the total of the payments made to the FPO by all the shareholders Members on their shares. It represents a form of member commitment to the group and it defines each member's stake in the group. In a FPO registered as a Producer Company it shall consist of only equity shares. Every FPO shall maintain a general reserve in every financial year, in addition to any reserve maintained by it, as may be specified in articles. In a case where the FPO does not have sufficient funds in any financial year for transfer to maintain the reserves as may be specified in articles, the contribution to the reserve shall be shared amongst the members in proportion to their patronage in the business of that FPO in that year.

Each FPO should prepare a balance sheet and profit and loss account of each financial year, which will be laid before the shareholders at the annual general body meeting. Every FPO has to file its Directors' Report, the audited balance sheet and profit and loss account along with the proceedings and the annual return with the Registrar within 60 days from the day on which the balance sheet and profit and loss account were laid before the members at the annual general meeting. Electronic filing is mandatory. As a part of annual filing, companies incorporated under the Companies Act, 1956 are required to file balance sheet, profit and loss account, annual return along with the e-Forms to the Registrar of Companies.

Financial requirements of FPO

Any business will have financial requirement to start the business and run the business. The capital requirement will depend on the nature and volume of business, which would vary from case to case. The cost will include both fixed and running cost.

Fixed cost

It is the expenditure, which will normally be a one-time expenditure. The expenses on the set up of the office with required furniture, fixtures and other equipment like computer, printer, cupboards, telephone connectivity etc. are fixed cost. Normally the FPOs are engaged in the activities of procurement, aggregation and grading of raw produce before sale. In such cases, Infrastructure like warehouse, weighing machine, graders/sorters, etc. will be required for any FPO which is also a fixed cost. This infrastructure can be purchased or can be taken on rent depending upon the situation.

Working capital requirement

The cost required by a FPO its day to business is called 'running cost' or working capital requirement. The working capital of any business unit is calculated based on the following criteria: (a) Procurement of Raw material, storage cost, processing, transportation, insurance (b) Management and administration cost for day to day activity of FPO, which may include staff salary, travel, rent, electricity, water,telecommunication, phone/fax, stationary, cleaning, meeting expenses of Board of Directors and the General Body, license fee, insurance and other statutory fee and other miscellaneous expenses (c) Training and capacity building of FPO functionaries: Training on subjects, like provisions in the Act, rules and regulations, statutory compliances, roles and responsibilities of Board of Directors and General body, banking operations, and also by exposure visits to the successful FPOs

Sources of Finance for FPO

The finance can be arranged from the following sources. (a) **Own Resources:** The reserve and surpluses of previous years are the source for personal financing. However, in case of a new FPO this opportunity will not be there. (b) **Suppliers' Credit and Advance Payment from**

Buyers: Suppliers' Credit can be obtained from credit companies or from potential buyers and sellers. The producers, who sell their products to the FPO, can sell on credit. FPO can get part payment in advance from prospective buyers. It can get agriculture inputs from the Agro dealers on the conditions of payment after sales. But mostly this type of finance is not available for start-up businesses or a new venture. (c) **Equity**: In case of a FPO the equity comes from the members and no external financier can participate in the equity investment. (d) **Grant support:** The FPO being a small holders' organization may seek capital support and other assistance from the Government under certain government schemes. Two major initiatives to support Farmer Producer Organisations (FPOs) (i) support to the equity base of FPOs by providing matching equity grants, and (ii) setting up of a Credit Guarantee Fund to provide cover to banks which advance loans to FPOs without collateral has been announced by GoI. Small Farmers' Agribusiness Consortium (SFAC) will implement the Schemes. Funding may also be available from the Department of Rural Development and Panchayats, Ministries of Agriculture and Cooperation or Horticulture or Food Processing GoI and or state Governments under various schemes like National and State Horticulture Mission, Small Agribusiness Consortium. World Bank, bilateral/ multilateral donor agencies and corporates under CSR may be other possible source of funds/grants from Producer organisations. The FPOs will have to develop a financially viable business plans for the purpose (d) **Debt financing:** This is the most preferred way of financing a new business. Here it is a direct obligation to pay the interest on the money lent by the financier. The biggest advantage is that the financier does not have control over the business as opposed to equity financing. The important point to be noted in this is the rate of interest charged. However, it is not easy to raise debt financing for a producers company without collateral and margin.

Governance of Farmer Producer Organizations

The administration or governance of a FPO is done by the various actors in the FPO such as the shareholders or the members, board of directors and the executive staff of the company or the office bearers. The general body of the company is the highest authority and the members of the FPO act through the General Body. The powers of the general body include,(a) Approving the annual budget of the company (b) Adopting the annual accounts of the company (c) Approving the quantum of withheld price (d) Approving the patronage bonus (e) Authorizing the issue of bonus shares to the members (f) Appointing an auditor for auditing of the annual accounts of the company (g) Declaring the dividend and deciding on the distribution of patronage (h) Effecting the amendments to the Memorandum of Association and Articles (i) Specifying the conditions and limits of loans that may be given by the Board to any Director (j) Approving any other matters that are specifically reserved in the Articles for decision by the Members.

Board of Directors: The Board of Directors is an important executive body of the company. Every company should have a Board of Directors. The number of directors should not be less than five and more than fifteen in number. The Board of Directors may act only in those areas that are not reserved to the General Body and may not exercise executive functions. In general, the board has the authority and is responsible for formulating, supervising, and monitoring of the performance of the FPO in respect of matters such as (1) Determination of the quantum of dividend payable to the members. (2) Determination of the quantum of withheld price and recommended patronage to be approved at the general body meeting. (3) Admission of new members to the company. (4) Pursue and formulate the organizational policy, objectives, establish long-term and annual objectives (5) Approve corporate strategies and financial plans of the company (6) Appointment of a CEO and other officers, as may be specified in the Articles and as per the requirement of the company (7) Exercise supervision, direction and control over CEO and other officers of the company. (8) Sanction any loan or advance, in connection with the business activities of the FPO to any member, not being a director or his relative (9) Investment of the funds of the company (10) Acquisition or disposal of property of the company (11) Exercise supervision over the book keeping especially 'books of account' of the company (12) Ensure the annual auditing of the company (13) Presenting the annual accounts in the annual general meeting with the auditor's report (14) Take such measures or do such other acts as may be required in the discharge of its functions or exercise of its powers.

Office bearers: The officers that are formally appointed by the company to look after the day-to-day affairs of the company include Chief Executive Officer and team. The Board of Directors from amongst persons other than members appoints Chief Executive Officer (CEO) on full time basis. The CEO shall be the ex-officio director of the board and shall not retire by rotation. The CEO shall be entrusted with substantial powers of management as may be determined by the board. CEO is accountable for the performance of the company, both, to the Board of Directors and to the members.

Policy Support from Central Government

The central government extends support to FPOs. The Department of Agriculture and Cooperation (DAC), Ministry of Agriculture& farmers Welfare act as the nodal agency for the development and growth of FPOs. The Small Farmers' Agribusiness Consortium (SFAC), A Society of DAC, is the designated agency of DAC to act a single-window for technical support, training needs, research and knowledge and to create linkages to investments, technology and markets. SFAC provides all- round support to State Governments, FPOs and other entities engaged in promotion and development of FPOs. In particular, SFAC is making efforts to create sustained linkages between FPOs and inputs suppliers, technology provides, extension and research agencies and marketing and processing players, both in the public and private sectors. The mandate of National Cooperative Corporation (NCDC) is being expanded to include FPOs in the list of eligible institutions, which receive support under the various programmes of the Corporation. NAFED takes steps to include FPOs in the list of eligible institutions, which act on its behalf to undertake price support purchase operations. DAC works with Food Corporation of India (FCI) and State Governments to encourage them to include FPOs as procurement agencies under the Minimum Support Price (MSP) procurement operations for various crops. DAC and its designed agencies also work with NABARD and other financial institutions to direct short term credit for working capital and infrastructure investments needs of FPOs. DAC also works with all relevant stakeholders to achieve 100% financial inclusion for members of FPOs and link them to Kisan Credit Cards. DAC also work with Ministry of Corporate Affairs and other stakeholders to further clarify and strengthen provisions of the law relating to the registration, management and regulations of FPOs with a view to fostering fast paced growth of FPOs.

Role of State Governments

The state government plays an important role in extending support to the producer's companies in different ways (1) By declaring the farmer's producer's organizations at par with cooperatives registered under the relevant state legislation and self-help groups/ federations for all benefits and facilities that are extended to member-owned institutions from time to time the state government supports FPOs (2) By making provisions for easy issue of licenses to the farmer's producer's organizations to trade in inputs (seed, fertilizer, farm machinery, pesticides etc.) for use of their members as well as routing the supply of agricultural inputs through farmer's producer's organizations at par with cooperatives (3) By using FPOs as producers of certified seed, saplings and other planting material and extending production and marketing subsidies on par with cooperatives (4) By suitable amendments in the APMC Act to allow direct sale of farm produce by farmer's producer's organizations at the farm gate, through farmer's producer's organizations owned procurement and marketing centers and for facilitating contract farming arrangements between farmer's producer's organizations and bulk buyers (5) By appointing FPOs as procurement agents for minimum support price operations for various crops. (6) By using farmer's producer's organizations as implementing agencies for various agricultural development programmes, especially RKVY, NFSM, ATMA etc. and extending the benefits of Central and State funded programmes in agriculture to members of farmer's producer's organizations on a preferential basis (7) By linking farmer's producer's organizations to financial institutions like cooperative banks, State Financial Corporations etc., for working capital, storage and processing infrastructure and other investments.

Initiative of Government of Karnataka for FPOs

Government of Karnataka has taken a special initiative for FPOs in Horticulture sector. Horticulture is an important sector in Karnataka, especially from the perspective of livelihoods, as it contributes more than 40 percent of the total agricultural GDP of the state. The Department of Horticulture and the University of Horticultural Sciences, Bagalkot have a significant role to play in the development and growth of horticulture in a state. As part of efforts to strengthen Integrated Horticultural Development through Public Private Partnership, Department of Horticulture, Government of Karnataka (DoH-GoK) has been promoting Farmer Producer Organisations (FPOs) as institutional vehicles for mainstreaming producers in horticultural value chains. The primary goal of FPO's is to provide services that support producers in their farming activities, including marketing of farm products. With support of various Resource Institutions (RI), DoH-GoK has promoted around 100 FPOs. It has taken proactive measures to empower the newly formed FPOs by providing a range of financial and non-financial assistance such as revolving fund, infrastructure development assistance, resource institution (RI) support for three years, risk fund and various other types of technical and institutional benefits. However, there is still a need to support FPOs for enabling their integration into horticultural value chains and mainstream markets. A Center of Excellence for FPOs (CoE-FPO) is registered and is being set up by the University of Horticultural Sciences, Bagalkot in its campus at GKVK Bengaluru. The university is also promoting FPOs in the state and helping them getting linked to the programs of the state government. The COE-FPO at UHS campus Bengaluru will handhold and guide the FPOs to act as effective business enterprises working for socio-economic advancement of their members.

The COE-FPO will act as a think tank to the Department of Horticulture and allied departments in the farm sector. The COE-FPO will engage in policy analysis, policy research, develops knowledge base within the state for gearing up the integrated development of farmers through FPOs. The activities of Department of Horticulture, University of Horticultural Sciences, other Development Departments and SAUs pertaining to collectivization of farmers will be attempted to be integrated by the COE-FPO and the results fed into the policy pertaining to farmer's development and welfare. It will also serve as a platform for coordination among various departments and other related government agencies, NGOs, private sector firms, farmers associations and other organizations dealing with the horticulture and allied sectors.

Objectives of COE - FPO

(i) To impart training and handholding assistance to FPOs for their development and evolution into effective enterprises working especially for economic benefit of smallholder members (ii) To become an institution of pre-eminence for research, learning and knowledge dissemination on Producer Enterprises in general and Farmer Producer Organisations in particular (iii) To act as an international think-tank for guiding policy and developmental interventions aimed at Producer Enterprises in general and Farmer Producer Organisations in particular (iv) To demonstrate an effective institutional model that promotes economic sustainability and self-reliance while serving the cause of disadvantaged entities such as Producer Enterprises and small holders.

It will also focus on (i) Knowledge base build up and knowledge management (ii) Transform the approach to water resources from that of development/construction to management (both reforming and performance) starting with the Irrigation and Drainage sub-sectors (iii) Establish and streamline coordination and integration among the various departments/agencies. (iv) Promote partnerships between the government and civil society inclusive of various stakeholders of horticulture and allied sectors.

Functions of COE-FPO

(i) to function as an advisory group agency for FPO's to co-ordinate policies and programmes related to sustainable FPO management (ii) to coordinate with Horticulture Department and other allied development departments, farm universities, in strengthening their respective core area of activities and enhance the impact of the outputs (iii) to refresh and enhance professional capabilities of CEOs of FPOs and other members and extension workers through Training, Consultancy and Research (iv) Collaboration with national institutions - both academic and research for bringing in new learnings to the state and partner for implementing best practices of FPOs.

References

1. Alur.A.S, Ravinder Reddy C.H, Gowda C.L.L, Reddy. B.V.S. Rai K.N, ParthasarathyRao. P, and Sunilkumar. S. 2008. *Building stronger farmers associations.* Global theme on Crop Improvement, International crops Research Institute for the Tropics. Pp. 32.

2. Camren K. and Keith, K. 1994. *Community consultation techniques: Purposes, Processes and Pit fall - A guide for planners and facilitators,* Brisbane, Department of primary industries, Queensland.

3. Chamala S. 1990. Establishing a Group: A Participative action model. In S. Chamala and P.D. Marten (Eds): *Working together for land care group management skills and strategies.* Brisbane, Australian Academy Press. Pp. 13-28.

4. Chamala S. 1991. Establishing a Group: A model for Participative action management. In P.D. Marten and S. Chamala(Eds):*Group Management skills for land care: A trainers guide*. Brisbane, Australian Academy Press. Pp. 33-60.

5. Nayak, Amar KJR. 2012b. Optimal Market Boundary with minimal Characteristic Distance between Small Producer and Customers: A strategy to realize higher value by both small producers and consumers from a transaction, *XIMB Sustainability Seminar Series,Working Paper 3.0*. 2012.

6. Nayak, Amar KJR. 2012c. Institutional & Organizational Asymmetries: Small Producers and Sustainability of Rural Agricultural Communities, *XIMB Sustainability Seminar Series 2.0, Working Paper*. 2012 and Keynote Address at the National Workshop on Markets that Empower Farmers (& Consumers), XIMB-ASHA, XIMB, July 30, 2012

7. *Policy and process guidelines for farmer producer organization*. Department of Agriculture and Cooperation, Ministry of Agriculture, Government of India. 2013.

❑❑❑

Chapter 9

Status and Prospects of Farmer Producer Organizations in Mysuru District of Karnataka

Arun Balamatti and Maheshwar K J

ICAR JSS KVK, Suttur, Mysuru

According to the Agriculture Census (2011) the total number of operational holdings in India numbered 138.35 million with an average size of 1.15 ha. of the total holdings, 85 per cent are in marginal and small farm categories of less than 2 ha. These small farms, though operating only on 44 per cent of land under cultivation, are the main providers of food and nutritional security to the nation. However, they have limited access to technology, inputs, credit, capital and markets. Small holders face several more constraints such as, small quantities of marketable surplus leading to poor bargaining power, scarcity of capital, lack of market access, limited knowledge and information, market imperfections, and poor infrastructure.

In order to ensure livelihood security to small and marginal farmers, it is necessary to focus on collectivization of farm holdings in India. A renewed interest in farmers' organizations has been growing in recent years with the government of India's policy intervention on farmers' producer companies. As a result, the farmers can now come together, exclusively, to work towards aggregation of their production process as well as input and output trade.

A Producer Company (PC) is relatively new legal entity of the producers of any kind, viz., agricultural produce, dairy and fish produce, forest produce, artisanal products, or any other local produce, where the members are primary producers. PC, as a legal entity, was enacted in 2003 as per the Section IX A of the Indian Companies Act 1956. Since the above enactment, the PC has been hailed as the organizational form that will empower and improve the bargaining power, net incomes, and quality of life of small and marginal farmers/ producers in India.

The farmers form the PC with the equity contribution. While each member in a PC can have only one vote, he/she can contribute different amounts of share capital to the PC. The day-to-day operations are expected to be managed by the experts, hired from outside, if necessary, under the direction of the Board of Directors elected/selected by the General body of the PC for a specific tenure. Since farmers or the primary producers are the equity holders of the company, a PC, as an organization, provides an appropriate framework for owning the company by the producers themselves. The need to organise farmers, especially the small holders, is a well-established fact. The basic purpose of the PC is to collectivise small farmers or producers for (a) backward linkage to mobilise inputs like seeds, fertilisers, credit, insurance, knowledge and extension services, and (b) forward linkages such as collective marketing, processing, market led agriculture production, etc., in order to get better share for farmers in the consumer rupee.

FPOs in Mysuru district

Mysuru district lies in the southern transition and dry zones of Karnataka. A wide variety of crops are cultivated in the district. Major field-crops are paddy, ragi, maize, black gram, green gram, horse gram, sugar cane, cotton and tobacco. The major horticultural crops grown include banana, vegetables (tomatoes, chilly, yard long bean, dolichos, brinjal and gourds), turmeric, ginger, and flower crops.

Mysuru district could well be the first to have started a PC, also called Farmer Producer Organization (FPO). MYRADA, a reputed non-government organization (NGO), registered the first FPO in Mysuru, Kabini Organic Farmers Producer Company Ltd., in the year 2010-11. The setting up of Kabini Organic FPCL was initially facilitated by the Dutch consulting organization ETC and the Rabo Bank. It is now being supported by NABARD. The FPO was started in the H.D. Kote block of Mysuru district by mobilising small and marginal farmers involved in organic production of cotton and vegetables.

Subsequently, many more FPOs are established in the district in the last 5 years, mainly on account of the efforts of NABARD and the Department of Horticulture and with the active involvement of different NGOs. As many as 12 FPOS are registered in the district wherein three FPOs have started trade activities whereas the rest 9 are in the formative stages.

The ICAR JSS Krishi Vigyan Kendra has made an attempt to look into the status and prospects of these FPOs by visiting each FPO, collecting the basic information using a semi-structured questionnaire, and by reviewing secondary literature. Here are the excerpts.

Major activities of FPOs

The essence of an FPO is to enable farmers to become 'farmer-entrepreneurs'. This, it is expected, to be attained by the farmer members pooling their resources to engage in collective production, input and output marketing. On these lines, the FPOs have started offering a wide variety of services to their members. All FPOs are trying to provide end-to-end services to their shareholders covering almost all aspects of cultivation such as supply of inputs, offering farm equipment on rental basis, and procuring and marketing of the members' farm produce. The FPOs are also facilitating linkages between farmers, processors, traders and retailers to coordinate supply and demand of inputs and farm outputs. Awareness creation on appropriate usage of fertilizers and distribution of soil health cards, with the help of KVK, are the technical initiatives common to the functional FPOs. The KVK is organizing customised technical training programmes to FPOs in crop production, integrated pest and disease management, integrated nutrient management, judicious use of irrigation water in field and horticultural crops. The FPOs seem to have better grasp of input procurement in bulk, especially the fertilizers. The produce marketing, though started by three FPOs, is still very limited in scale and frequency.

An overview of 9 out of 12 FPOs registered in the district is given below.

Table 1. Status of FPOs in Mysuru

	FPOs							
	Kabini Organic FPCL*	**Raitha Mitra FPCL**	**Econut FPCL**	**Kavery Rural Devt. Coop. Soc.**	**Vandana Mahila Shangha**	**Hussai-npur Hort. FPCL**	**Varuna Hort. FPCL**	**Nanjan-gud Hort. FPCL**
Promoter	MYKAPS	NABARD	CDB, NABARD	MYKAPS	MYKAPS	DoH & SFAC	DoH & SFAC	DoH & SFAC
Year of est.	2010	2015	2015	2016	2015	2016	2016	2016
BoDs	5	10	8	11	11	12	12	10
Shareholders	1302	526	548	872	207	720	890	1000
Authorised share capital	25,00,000	19,90,000	5,00,00,000	5,00,000	3,00,000	10,00,000	9,00,000	10,00,000
Paid up capital	10,00,000	10,00,000	10,10,000	47,100	1,45,800	15,00,000	1000000	1000000
Turnover	3,40,43,659	2,11,000,00	50,00,000	4,60,000	3,86,660	8,55,000	3000000	3500000
Annual profits	4,37,533	3,00,000	20,000	41,810	19,333	16,490	0	25,000

* Data for FY 2014-15 and the data for all other FPOs is for the FY 2016-17

The Table 1 implies that except Kabini Organic FPCL all other FPOs are rather new, established in 2015 and 2016. Only the Kabini Organic FPCL and Nanjangud Horticulture FPCL all other FPOs are yet to reach full membership. However, the Raitha Mitra FPCL has offered institutional membership to 50 Farmer Interest Groups (FIGs), each having 15-20 members, by virtue of which they have been able to enrol enough members and mobilise full share capital. The 2 FPOs, Kabini Organic and Raitha Mitra FPCL have crossed annual turnover in excess of Rs. 3 and 2 crores, respectively whereas the others are far behind. The annual profits indicate that the FPOs, despite doing transactions in large amounts, are not making big profits largely on account of initial capital investments, high operational costs and possibly investing on activities of low margins.

A SWOT analysis of the FPOs of Mysuru district is presented below.

Strength of FPOs

A member base of 1000 farmers, the paid-up share capital and the eagerness to move beyond production process and get involved in input and output trade are supposedly the strengths of any FPO. However, each FPO seems to have found its own unique strength. For instance, the Kabini Organic FPCL is focusing on organic farming, which has a growing market in Mysuru with demand for products of organic origin distributed around the year. The Raitha Mitra FPO, on the other hand, seems to be riding on the strength of its leader; the chairman of the FPO is the President of the Karnataka State Sugarcane Growers' Association and has a good following among the farmers of not only Mysuru but also in many districts of Karnataka state. Both these FPOs are also aggressively promoting cultivation and export of super food crop viz., Chia. Chia is now being cultivated on over 2000 acres in the district, thanks to the introduction by Kabini Organics FPCL and systematic promotion by Raitha Mitra FPCL. The Econut FPO, initially supported by the Coconut Development Board (CDB), is banking on another niche market, which is selling of fresh tender coconuts directly to IT professionals in Bengaluru by opening brand outlets right in the campuses of IT companies. This way, the FPO is able to procure nuts from its member farmers directly by paying one rupee more on each nut over and above market price and selling farm fresh nuts directly to consumers thus eliminating middlemen. The FPOs, thus, are banking on their respective intrinsic strengths.

Weaknesses

Member enrolment, weak sense of ownership among the members and lack of role clarity among the board directors seem to be the expected early stage weaknesses of FPOs. The FPOs in Mysuru are no different but there are also exceptions in that, the Kabini Organic FPCL is backed by the NGO, MYRADA, known for its strong community mobilization skills, and the Raitha Mitra FPCL has the advantage of strong leader who could garner members in quick time through his individual efforts. Procedural constraints and delays at the time of registration, bureaucratic delays in obtaining licenses for input trade, weak linkages and lack of trade experiences are also common constraints of the FPOs. The most important weakness, however, is that the FPOs, almost all of them, seem to be trapped in the cooperative mindset. Their activities largely appear to have been driven by the welfare motive rather than getting involved in such activities which are likely to fetch them high profits. Investments in purchase and distribution of chemical fertilizers is a testimony to 'high investment-low returns' activity, surprisingly common to all FPOs, whereas the FPOs seem to have no clue on exorbitant pricing of chemical as well as biological pesticides where they have hardly intervened. Poor capabilities of value chain analysis, business plan preparation and choosing lucrative commercial opportunities are the weakness which, unless the FPOs overcome quickly, could become serious deterrents for the FPOs. Lack of capital and infrastructure are the other common weaknesses.

Opportunities

A wide range of crops grown in the district, especially the crops that are commercially important, could be the immediate opportunities for FPOs. There are many crops that have lucrative business potential in the district in terms of volume of inputs traded and the nature of outputs that have niche market not only in Mysuru but even in the international market. For instance, Mysuru is known for cultivation of Virginia Flue Cured Tobacco which has a thriving global market. The district is also home to long staple cotton cultivation, again something that is much sought after in the international market. Being next to metro city like Bengaluru and state like Kerala, Mysuru has grown into a vegetable hub. Cultivation of fruits like banana and water melon offer large commercial opportunities. Because of cultivation of commercially important crops the FPOs have many opportunities in input trade as well as value addition and marketing of various farm produce. The question is, whether they are well placed to harness these opportunities. It is heartening to know that the Raitha Mitra FPCL has already moved forward in this direction and has started procuring vegetables produced by its members and is selling it to government outlets in Kerala thus avoiding middlemen.

Threats

Threat normally refers to factors that are beyond the control of FPOs. As such, the expectation from the government while passing the Act on FPOs is that the farmers should graduate from being mere farmers to become 'farmer entrepreneurs'. This idea emanates from the fact that despite the use of best of the technologies there is a limit to enhancing the productivity; the resultant increase in total production often leads to glut and price crash. Therefore, if the farmers were to increase their total profits they should also explore possibilities of reducing their production cost - by harnessing inputs at lower prices, and enhance the value of their produce by engaging themselves in value addition and or any other market functions. This is where the farmers need to be acquainted with the entrepreneurial skills. Unfortunately, that has exactly been the weakness of farmers, in general. The farmers have difficulty imbibing entrepreneurial abilities on their own and hence it is for the agencies promoting FPOs to bring in these skills. However, it is largely the NGOs which are promoting FPOs and the NGOs are also not known to be strong on entrepreneurial skills. Therefore, the farmers mobilised into FPOs are likely to be intimidated by the market forces, especially in the early formative stages. The climate change and the associated production fluctuations are making the markets that much more unpredictable, not an easy challenge for FPOs to handle in their infancy. The enigmatic markets and climate change are the biggest threats the FPOs will have to learn to deal with if they were to succeed.

Conclusion

Mysuru district can take pride in having started the first FPO in Karnataka viz., Kabini Organic FPCL. The Kabini Organic FPCL and the Raitha Mitra FPCL have introduced a new crop, the super food Chia, as part of bringing more revenue to their farmer members, and have been largely successful in doing so. At the same time, except for the Kabini Organic FPCL promoted with a clear purpose of promoting organic farming among its small and marginal farmers and the leader-driven Raitha Mitra FPCL, the rest of the FPOs seem to have cropped up more as a means of meeting the target, promoted mostly by external agencies and not from the farmers' own initiative. It is evident from the examples of the FPOs, though quite young, and their trade experiences so far, that there is a need to strengthen them with training and handholding support on value chain analysis, business plan preparation and identification of core competencies of each FPO and building appropriate commercial activities. The success of couple of FPOs is expected to infuse energy and enthusiasm among the rest of the FPOs. The long-term success, however, depends as much on the farmers' own willingness to grab this opportunity as with the earnestness and competencies of the promoting agencies.

Chapter 10

Onion Seed Production for Self Reliance and Higher Returns

Kerure P, Channagouda R F, Onkarappa S and Bindu B M

ICAR - KVK, Babbur Farm, Hiriyur, Chitradurga

Chitradurga district is located in the Central Dry Zone of Karnataka (Zone IV) with average rainfall of 450-500 mm. The major crops of the district are finger millet, maize, redgarm, groundnut, cotton, onion, coconut, banana, arecanut and millets. The district average yield of onion is 19.50 t/ha but, potential yield is around 30 t/ha, which indicates that there is scope for increasing the productivity and returns. The reasons for lower productivity are poor quality seed material, imbalanced nutrient management and lack of awareness on plant protection measures. Based on the problems observed, KVK Chitradurga started conducting on farm testing on nutrient management for addressing the issues related to leaf twisting which was one of the major problems for yield decline. Later, large scale frontline demonstrations were conducted on integrated crop management in onion in Chitradurga district and several seminars and trainings were conducted to create awareness and spread the technology. From past five years the KVK started seed production activity in onion through FLD by farmer's participation. Since, it is mainly grown for commercial bulb production during *Kharif,* very less number of farmers growing seed production during *Rabi* season with local varieties. In this contest, KVK planned for technical interventions on scientific seed productions and quality seeds in district.

Scenario of the district and situation analysis

The area of onion in *kharif* is around19,193 ha accounting for around 2 lakh kg seed requirement. The yields of onion vary from 15 to 20 t/ha and it has been demonstrated through front line demonstrations that onion bulb yields could be increased by adopting

ICM practices and Arka Kalyan variety of IIHR Bengaluru, which is suitable for this region. Seed is the major input for which farmers are dependent on Satara Dist. of Maharastra. But quality is not assured and there are many instances of susceptibility of the variety for pest and disease. Hence, it was felt necessary to train the farmers on scientific seed production technology to ensure seed availability in the district.

Strategies adopted by KVK

Based on the constraints faced by the farmers and problems identified by KVK, it was planned to take up quality seed production through front line demonstrations, trainings and organise field days for wider dissemination of the technology (**Fig. 1**).

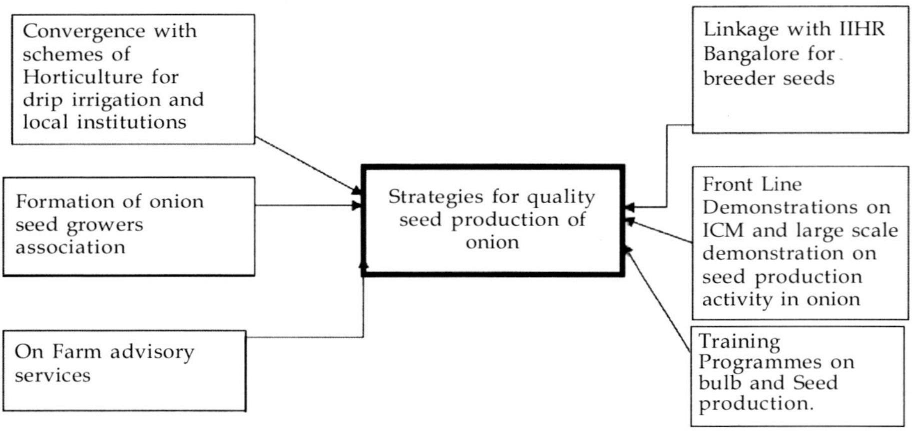

Fig. 1. Interventions of KVK

Organisation of sensitization and training programmes

KVK started assessment of different suitable technologies for correcting the leaf twisting problem during 2008-09. The results revealed that application of zinc, boron and gypsum reduced the percentage of leaf twisting in onion. In the succeeding years demonstrations were conducted on integrated crop management to improve the bulb yield and during 2012-13 onwards were given more emphasis on seed production activity in major onion growing areas of Chitradurga. Apart from importance of seed production and its economics were convinced through district level seminar, workshops and krishi melas. The training programmes on ICM, nutrient management, weed management and scientific ways of seed production activity were conducted through FLD's and Farmers Field School (Table 1 and Fig. 1).

Front line demonstrations

Onion is major traditional crop of our Chitradurga district and initially KVK introduced the Arka Kalyan promising variety of onion in three villages of Hiriyur by conducting large scale front demonstrations in 80 farmer's fields. Since, 2009 to 2016 technologies on bulb and seed production activities are being taken up in different cluster villages of Hosadurga, Holalkere and Hiriyur talukas. Farmers were educated on seed treatment, balanced nutrient management and pest and disease management. Later on the FLD's on integrated crop management in onion were conducted in, 2012-13 and 2013-14 by covering 10 ha in 25 farmer's field. After that KVK initially started seed production activity in improved varieties viz., Arka Kalyan and Bhima Super during 2013-15, 2015-16 and 2016-17 in 15 ha through farmer's participation involving 15 farmers.

Large scale demonstrations under 'Seed Village Programme'

In order to ensure quality seed availability in the district, seed production activity was started with Muddapura village during 2012-13 with eight farmers and around 20 q seed was produced. This seed were spread through KVK by providing market linkage. Seed was spread to neighbouring districts viz., Bellary, Chikkamagaluru and Tumkur (Table 2 and Fig. 2).

During 2016-17 under UAHS, Shivamogga an innovative programme on participatory seed production activity was taken up to evolve "Seed Village Programme". Fifty farmers of Hosakundur village of Hosadurga taluka, Ramagiri village of Holakere taluka and Babbur of Hiriyur taluka took up the seed production activities in 50 acres. Breeder seeds of Arka Kalyan, Bio inputs, Yellow Sticky trap as a part of IPM kit, Bee hives were given to farmers and various on campuses, off-campus trainings, field visits and filed days were made to the demo plot and various advisory services on plant protection and nutrient management were given.

On farm testing

Varietal assessment was made to know the performance of different varieties during 2013-14 and 2014-15. Among them Bhima super and Arka Kalyan were found to be promising, and this outcome helped to take up large scale seed production activity programme during 2016-17.

Farmer Field School programme

Farmer Field School was conducted in Basappana Malige, Vaddikere, Beerenehally, Burujinroppa, during 2011-12 under NAIP project, large scale demonstrations were taken up in Basappana Malige, MD Kote and Vaddikere village of Chitradurga. The Farmers field school helped the farmers to understand the relation between biotic and abiotic factors, and reduces the cost on fertilizers, pesticides and improved the yield by reducing the cost of production and protecting the crop ecosystem.

Farm advisory services

The various advisory services on variety, nutrition, weed management, micro irrigation, integrated pest and disease management, seed production aspects viz, isolation distance, rouging, role of pollinators, seed extraction and storages, packages and formation seed grower association.

Convergence with line departments/organisations

Convergence and linkages with the schemes of line departments and organisations involved in seed production activity, providing drip irrigation system through linkages with State Horticulture Departments, National Horticultural Mission and Mahatma Gandhi National Rural Employment Guarantee Programme (MGNREGP) and NABARD. KVK also developed linkages with Non-Government organisations viz., MYRADA, and GRAMA for promotion of seed production activity by technical support and farm advisories.

Table 1. Field extension activities and number of farmers involved

S.No.	Programmes	No. of programmes	No. of farmers
1.	Training programmes	69	1964
2.	Front Line demonstrations	8	80
3.	OFT	4	12
4.	Extension activities	105	2674

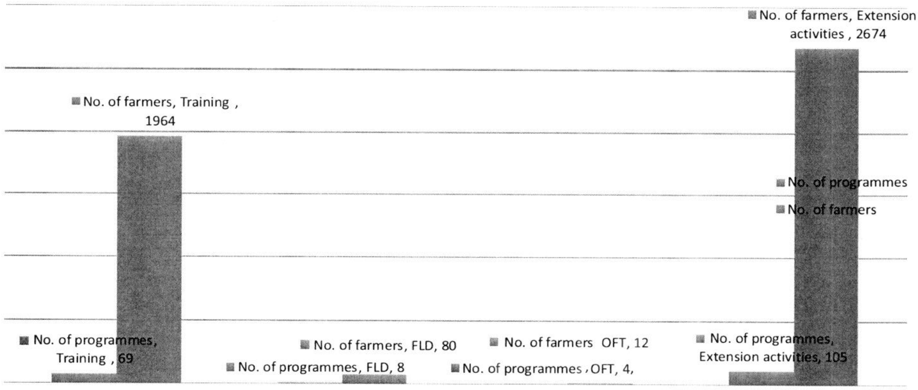

Fig. 2. Field extension activities and number of farmers involved

Table 2. The year wise seed production of onion variety Arka Kalyan and No. of farmers involved

Sl. No.	Year	Seed production (q)	No. of farmers
1.	2012-13	12	8
2.	2013-14	20	8
3.	2014-15	22	10
4.	2015-16	26	12
5.	2016-17	192	50
	Total	**272**	**88**

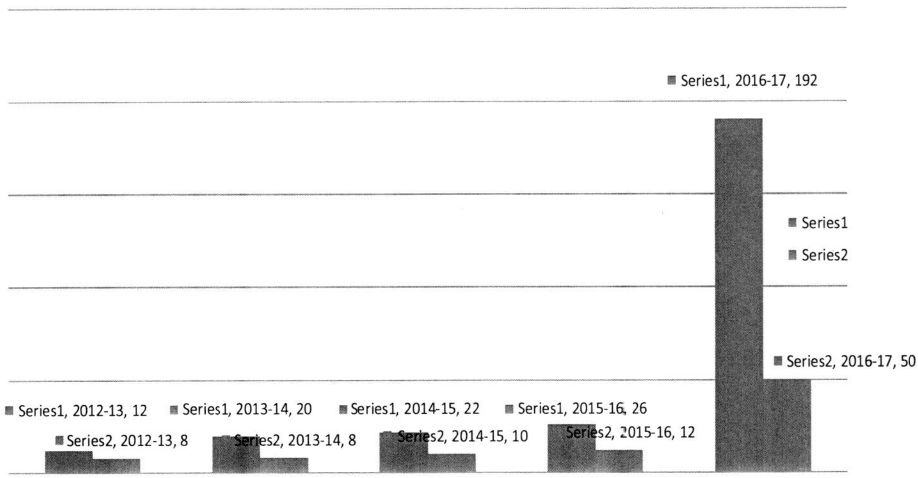

Fig. 3. The year wise seed production of onion variety Arka Kalyan and No. of farmers involved

Outcome and impact of interventions

1. Adoption of onion variety Arka Kalyan and area expansion

Comprehensive interventions of KVK through awareness programmes, training activities, Front Line Demonstrations, advisory services, seed production and convergence with line departments has resulted into a positive outcome in terms of adoption onion variety Arka Kalyan by the farmers of the district. There has been a significant increase in area under onion variety Arka Kalyan. The details of area spread in the district are presented in Table 3 and Fig.3.

Table 3. Year wise Expansion of Area under onion variety Arka Kalyan in the District

S.No.	Year	Seed production (q)	Area Expansion (ha)
1.	2012-13	12	120
2.	2013-14	20	200
3.	2014-15	22	220
4.	2015-16	26	260
5.	2016-17	192	1,920
	Total	**272**	**2,720**

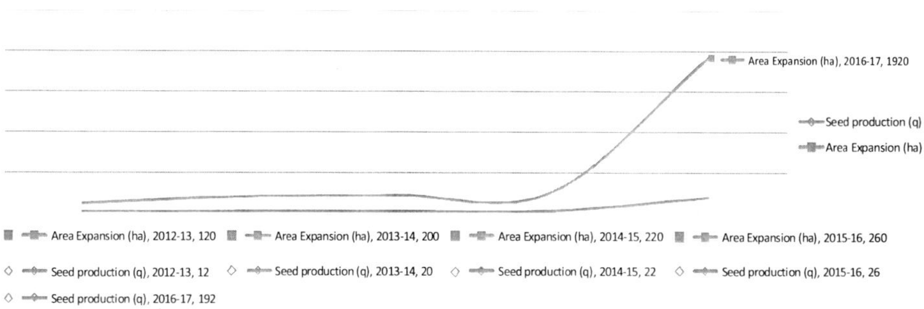

Fig. 4. Year wise Expansion of Area under onion variety Arka Kalyan in the District

The information depicted in figure reveals that during the period from 2012-13 to 2016-17 the area under onion variety Arka Kalyanis spread to the extent of 2,720 hectares involving 272 farmers. When this data is compared with benchmark data of the year 2006-07, there has been 172 percent increase in area over the intervention (Table3 and Fig. 3).

2. Economics of onion commercial seed and bulb production

Onion cultivation is economically viable for the farmers as compared to local variety which is mainly cultivated in Chitradurga district. Onion variety Arka Kalyangives returns even during agricultural drought years and resistant to purple blotch. This is in contrast to the local varieties cultivated by the farmers.

The comparison of economics between commercial seed and bulb production of Arka Kalyan variety presented in Table 4 and Fig.4. The results (FLD data of 2015-16) revealed that commercial seed production activity gave high net returns of Rs. 7, 10,000.00 compared with commercial bulb production of Rs. 2, 05,000.00.

Table 4. Economics of onion commercial seed and bulb productions

S.No.	Onion varieties	Average yield (t/ha)	Cost of cultivation (Rs./ha)	Gross returns (Rs./ha)	Net Returns (Rs./ha)	B:C ratio
1.	Arka Kalyan (Commercial Bulb production)	30.0	81,378	2,86,378	2,05,000	3.51
2.	Arka Kalyan (Commercial Seed production)	1.25	1,65,000	8,75,000	7,10,000	5.30

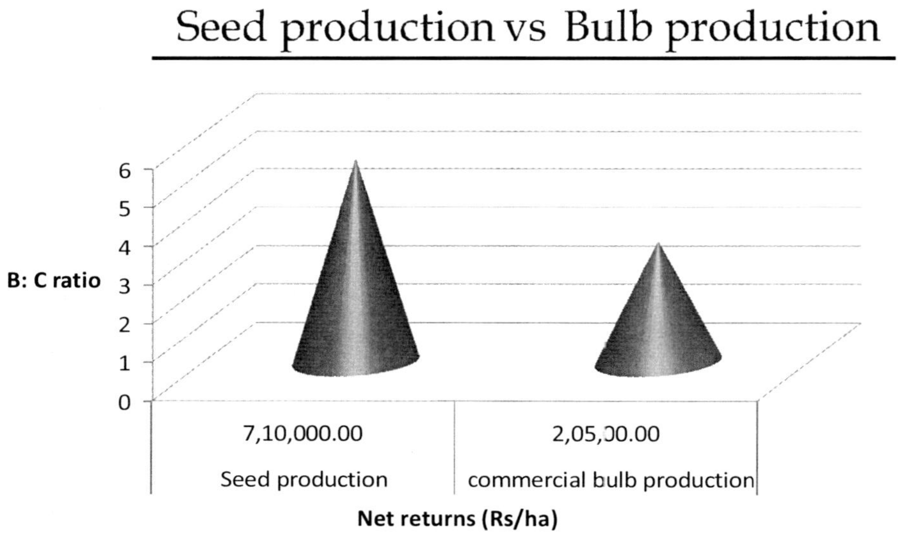

Source: KVK Chitradurga, Annual Report2015-16

Fig. 5. Economics of onion commercial seed and bulb productions

Conclusion

The production of onion gave higher net returns compared to commercial bulb production and also provides additional employment generation to family members of farmer. It is helps to farmers of Chitradurga district for self reliant of quality seeds and timely availability at their local region by reducing the risk to purchase of seeds from neighbouring state.

❑❑❑

Chapter 11

Puthari Farmers Producers Company Limited (PFPCL), Gonikoppal, Kodagu, Karnataka: A KVK - Facilitated FPO

Saju George* and Venkat Kumar R**

* ICAR-KVK, Gonikoppal, Kodagu and ** ICAR-IIHR, Bengaluru

A cup of coffee brings a lot of happiness to people. It is one of the most preferred drinks in the world. Kodagu district is famous for its coffee and black pepper which is mostly shade grown and the most environment friendly coffee and pepper grown anywhere in the world.

But nowadays it has become very challenging for the farmers to grow coffee and pepper economically because of various issues faced by them. Labour problem, sky rocketing farm input prices, increase in plant disease/pests, falling prices of the end product and to add to it the manipulation/malpractices of traders and agents. When will a coffee grower get a new lease of life? This is the same question, which kept bothering coffee growers in a small town called Gonikoppal, located in Kodagu, which is known as the Scotland of India.

What was the answer then? To answer this question, came help from Krishi Vigyan Kendra scientist, who suggested these farmers to start a company on their own. Farmers Producer Organization (FPO) was born. This new seed came to be known as Puthari Farmers Producer Organization, "Puthari" meaning new rice and also a paddy harvest festival of Kodava community, which is celebrated with all their heart and soul put into it. This became a "Ray of hope" for the grower who had lost all hope in the past.

Puthari Farmers producer company Ltd (PFPC), took roots by the help of financial support from NABARD, Kodagu and ICAR-KVK, Gonikoppal became its stem with their technical guidance and company initially started by 27 of the fellow farmers became its own branches. PFPC started building trust and confidence from the fellow coffee growers, even before it got registered. Keeping a motto of "Provide every farm input at wholesale price and buy every farm output of the estate at retail price" there by helping in increasing his income, sustainability and attracting young generation to the profession of Agriculture.

Puthari Farmers Producers Company Limited (PFPCL) is a Farmers Producers Company (FPO) facilitated by a Krishi Vigyan Kendra (KVK). The ICAR-Indian Institute of Horticultural Research (IIHR) KVK at Gonikoppal, South Kodagu district of Karnataka caters to the technology dissemination needs of farmers of this district. As a novel initiative to fulfill the motto of Hon'ble Prime Minister of India, Shri. Narendra Modi to double the income of farmers by 2022, through collectivizing farmers' action in agricultural supply and value chain activities, the KVK facilitated establishment of such FPO in active collaboration with National Bank for Agriculture and Rural Development (NABARD), Kodagu district, Karnataka.

Resource organizations of the FPO

- ICAR-IIHR KVK, Gonikoppal, Kodagu district
- National Bank for Agriculture and Rural Development (NABARD), Kodagu district

Producer organization promoting institution (POPI)

ICAR-IIHR KVK, Gonikoppal, South Kodagu district technically nurtures this company for

- Mobilizing farmers
- Registration
- Appointment of office-bearers
- Business planning and networking
- Technological backstopping/ agro-advisory
- Capacity building of office-bearers and members
- Advisory on agricultural supply and value-chain activities

Objective of the FPO

- To bring-down agricultural operational cost through attaining scale of economy
- Collectivizing farmers action in agricultural operations

Business activities proposed

- Organizing scientific platforms for production and marketing of black pepper in collaboration with Spices Board
- Organizing interfaces on Avocado post-harvest practices and marketing
- Distribution of farm inputs, tools and small implements on whole sale rates
- Reaching out many farmers in Kodagu district

A brief profile of FPO

This FPO was registered in early 2017. The members of FPO are the cultivators of coffee, pepper, coconut and seasonal fruits such as avocado, garcinia, sapota etc. The FPO has started functioning very recently and hence, the membership is gradually increasing. Presently, there are about 100 members, including 10 women members. The members represent 25 villages of Kodagu district of Karnataka. The share capital amount is Rs. 5000/share. The FPO collects an amount of Rs. 200/-per member as processing fee. The members, with an upper limit of 10shares/member, should purchase a minimum of 5 shares. As per the policy of the FPO, it will be one voting right for each member. The office-bearers of the FPO are consulting with various other FPOs to formulate their own bye-laws. The FPO plans to conduct election/ selection of Board of Directors and other office-bearers once in a year. The benefit sharing process, business performance details, procedures for following performance and auditing record keeping etc., are yet to be finalized. KVK, Gonikoppal and NABARD, Kodagu district are facilitating the office-bearers of FPO on these lines. Similarly, the company has to create its own infrastructural facility for office space, go-down, warehousing facility etc. Detailed plans are being prepared by the office-bearers.

Organizational structure

Being a young FPO, the organizational structure of the company is very simple as follows:

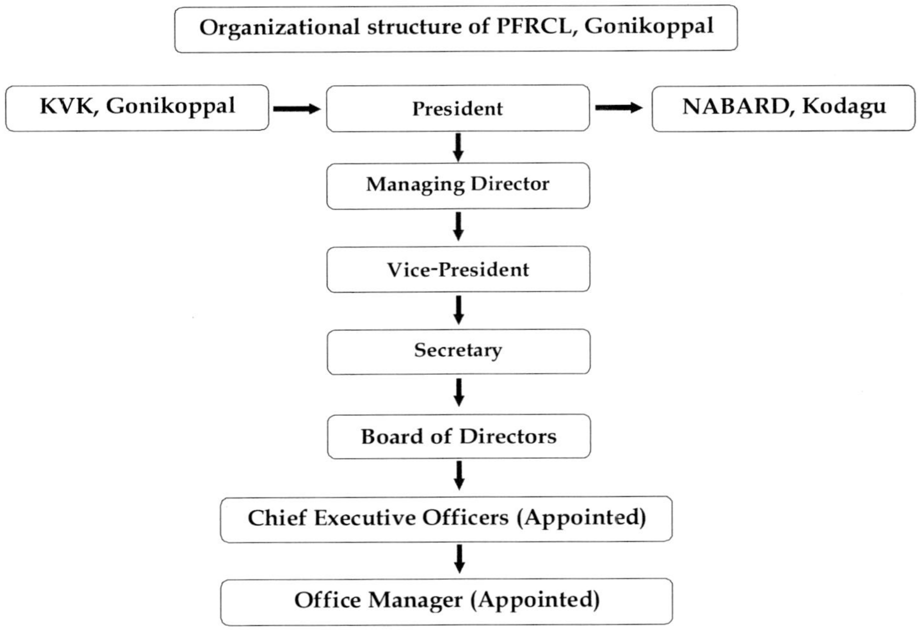

Input-supply services offered

- Sale of fertilizers, plant protection chemicals, micro-nutrients, bio-pesticides on whole sale prices (The FPO is trying to get licenses of such technologies developed ICAR institutes)
- Sale of seeds at whole sale price
- Sale of farm tools and implements
- Sale of kitchen garden kits

Procurement and marketing services

- Procurement of Avocado
- Marketing of Avocado through tie-up with private company Farm Folks

Technological advisory/services

- Capacity building for members through ICAR-KVK, Gonikoppal, ICAR-IISR, Appangala, ICAR-IIHR, Bengaluru
- Arranging need-based agro-advisory for members through ICAR-KVK, Gonikoppal, ICAR-IISR, Appangala, ICAR-IIHR, Bengaluru
- Purchase of technologies from ICAR institutes and arranging for production and supply of inputs such as micronutrients, bio-fertilizers and bio-pesticides

Critical success factors

- Supply of inputs on wholesale price
- Procurement and marketing of seasonal fruits such as avocado
- Transparency in all technical and financial dealing
- No political interference
- Adequate support from ICAR-KVK, Gonikoppal, ICAR-IIHR and ICAR-IISR

Constraints faced

- Slow membership drive
- Lack of adequate ideas towards dovetailing of Govt., schemes
- Lack of adequate ideas towards liasoning with financial institutions for credit and insurance for members

Future plans of FPO

- Value addition and branding for coffee and pepper
- Procurement and marketing facilitation for all seasonal fruits
- Facilitating effective input supply chain for members
- Establishing a mini-processing plant for coffee, pepper and seasonal fruits
- Establishing adequate infrastructure, especially warehousing facility

Conclusion

The Puthari Farmers producer company has seen good response and grown rapidly since its registration in March 8th 2017.Today this company has about 175 coffee growers as its shareholders and has aim of reaching 1,000 members within the next one year. PFPC has already procured all necessary licensing for its functioning; to name a few: Fertilizer/chemical, commodity trading license, FSSAI license and has opened a retail out-let to supply estate inputs and is in the process of Value Addition for products. PFPC is on its way to becoming the success story of a company which is run by the farmer's themselves. The vision is very clear and goals are set by the company to help coffee and pepper grower in every possible way, so that the farmer does not feel let down and start looking at this profession as the best work in his life. A farmer can take pride in being in a profession, which grows and takes care of "living beings" and also as a person who "feeds the entire world".

❑❑❑

Chapter 12

Institutionalizing FPOs through M4 Model for Sustainable Agri-business

Dhanakumar V G

IIPM, Bengaluru

Agrarian structure at the grassroots level is not a homogenous or harmonious entity and basic contradictions are a fundamental fact of grassroots life. According to the World Bank, the importance in sustainable development has lead development agencies to make grassroots institutional sustainability itself, a consideration in the planning of design. The shortage of sustainable and business-oriented homogenous groups/institutions at the grassroots remains a serious issue. The reasons for these shortages, and ways to alleviate it, are poorly understood with reference to (Farmers Producers' Organization (FPOs). Institution building means "planning, structuring, and guidance of new or reconstituted organizations that embody changes in values, functions, physical and/or social, environmental, technologies etc.," This article describes the analytical framework pertains to producers' collective for commodity business development, organizational structure, functional modules, people participatory processes and management strategy.

Institution is a subtle concept and therefore subject to confusion. It refers to stable, valued, recurring pattern of behaviour. Institutions thus include rules or procedures that shape how people act, and role or organizations that have attained special status or legitimacy. Sustainable Institution in a strict sense the term is redundant since institutions are by definition, sustained ways that people interact. But in development circles, the conventional meaning of this refers to consciously designed organizations that do one or more of the following: 1. they survive over time as identifiable units, 2. They recover some or even all their costs and 3. They supply a continuing stream of benefits to the group/community.

Producing a good fit between producers' collective and its internal and external situation is the task of strategic business-led group formation and management. Although it emerged first in private, profit-making sector of the industrialized countries, strategic producers' group management is now gaining attention in the public and non profit sectors. The underlying assumption is that there is more than one effective strategy for development institutions; the right one for any particular situation depends on the task, the environment and the stage of the organization's life cycle.

The fundamental theme of FPO, as institutional building is documented as a voluntary, self-initiated homogenous group, valuing personal interaction and mutual help/aid as a means of alluring or a meliorating the problems perceived as alterable, pressing and personal by most of its participants. These groups are voluntary togetherness of people formed to attain certain collective goals that would be economic, social, occupational (or) combined forms of need. And, a collective often helps members cope with their feelings and get group support. It also offers emotional support and practical help with a problem that is common to all members.

Process of Building Producers' Collective

Grassroots participation in development means consulting and involving the producers' in the identification of local needs, and conception, formulation and implementation of any project in order to develop the necessary Self–Reliance and Confidence (SRC) needed among the producers for accelerated agri-business and market development.

Participation is made possible through FPOs in which among other things, leaders emerge or are chosen from within the community by selection. The success of any FPOs depends on the operations of such leader. The success of business development programme does not depend on material resources alone. The human element is equally important. The leader, known to and having influence with those whose life he wishes to improve can indicate the direction of a proposed change, make clear objectives and values, and involve those it affects in bringing it about.

Process of Voluntarism for Producers' Collective

Producers' collective should develop many opportunities to strengthen local institution and improve the quality and relevance of services by drawing on private enterprises, NGO's, academic institutions, association of local producers' etc., for the improvement and performance of extension agents. There are many combinations available to strengthen local action capabilities. Instead of relying primarily on inadequate members of professionals and sub-professionals, development agents should advocate less formally trained and much less costly, local persons to extend extension service to grassroots development. The development organizations have already to support, project that employ paraprofessional or indigenous non-civil service personnel to improve the cost effectiveness through paraprofessional staffing. Paraprofessionals selected by, compensate by, accountable to producers' collective can increase the relevance and cost effectiveness of services while enhancing local self-reliance. Paraprofessional's role, function and skills as frontline service become increasingly the responsibility of producers' collective and development.

The purpose of this section is to explore strategic producers' homogenous group management methods that can increase the cost effectiveness of government expenditures in grassroots sectors through innovative staffing (volunteers) patterns, using local, service personnel or paraprofessionals do involve a modest operating cost as compared with cost of normal civil service staff. According to Cornell University working group, paraprofessionals are persons of indigenous, to the service area with the limited formal schooling and limited training. The para-professionals works under the overall guidance of extension agents. The sub professionals are men (occasionally women) who are graduates from secondary school and 2-3 years of technical education in institutes or colleges. Normally they have full time and personable public functionaries and related persons. For e.g. sub professionals are rural leaders, eminent youth, progressive producers' etc., who are in direct with the members of producers' collective organization. The important distinction is community paraprofessionals are responsible and accountable to the neighbors, not to bureaucratic agency.

Formation of Cluster-led-FPOs

As indicated earlier, formation of cluster-led-FPOs and its network have emerged as a strategic producers' participatory management approach to promote multifunctional interest of homogenous group, society/cooperative, federation and company. It is necessary that facilitating organization/NGO/development departments may give an adequate time for popular participation of members to understand the core objective of cluster while forming 4M stages of FPOs. In addition, care should also be taken with the basic principles, functional and structural dimensions of producers' organization, in the process of structural construction and building FPOs.

The process of formation of FPO cluster and its approach must focus on at M4 dimensions i.e. 'Micro (M1)', 'Meso (M2)', 'Macro (M3)' and 'Mega (M4)' levels. Existing traditional community structures, traditional institutions, organisation and community welfare organisations are effectively utilised to build trust and confidence of farmers at the micro-level (M1). Mobilisation of traditional leaders and involving them in community development is essential. At the same time, community based self-help structures are introduced and developed from the base up. Both participation and mainstreaming are taken as essential components for building internal development networks. Producers' cluster at the meso-level (M2) would help the community to reach the second level of development structure and external support systems with market network for cluster support as well as articulation of need and influencing policy and decision making. The third component of cluster process is at macro (M3) level, which facilitates the development activities as federation to link with market, government, commodity boards and other associated stakeholder sectors to jointly initiate the business process from bottom of (M1) micro level to meso (M3) sub-system and mega (M4) segment. These three (M1 to M3) levels of development concept for institutionalising cluster process and its link with M4 mega level are illustrated in Fig. 1.

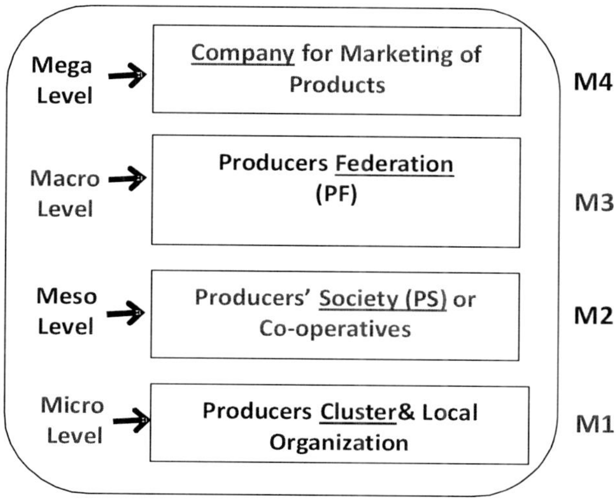

Fig. 1. Four levels of development concept for institutionalising cluster-led building process

The above section explains on how a producers' collective development programme at the grassroots must plan and develop a cluster of M4 as micro, meso, macro and mega levels in different phases and how to merge them together with sectors (e.g., Coconut, organic farming, arecanut, tea, coffee, etc.). An illustrative triangle model on formation of coconut sector with reference to M1 to M4 guidelines on formation of group is illustrative below in Fig. 2.

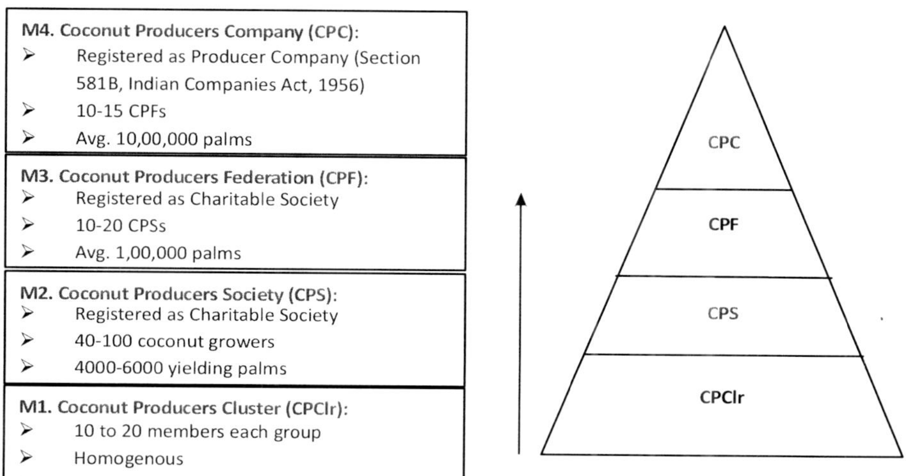

Courtesy: CCB Bengaluru

Fig. 2. Formation of FPOs: 4M structure coconut perspective

Therefore, Fig. 3 gives a holistic step-wise formation of M4 (refer to picture format) for producers' cluster with special reference to coconut sector on how to institutionalizing stakeholders' network for development. It also shows on how local, district, state and national level structures can function together in a perfect harmony within a sector, such as coconut. The four-stage process of producers' cluster at meso-level starts with a simple approach of facilitating participatory linkage derived from the base of micro-level. However, organising an interest group around "producers' interest" alone is not enough, because interest of the group must also extend to environment, group marketing, utilisation of community resources etc., as they are essential for sustaining the objectives of the cluster.

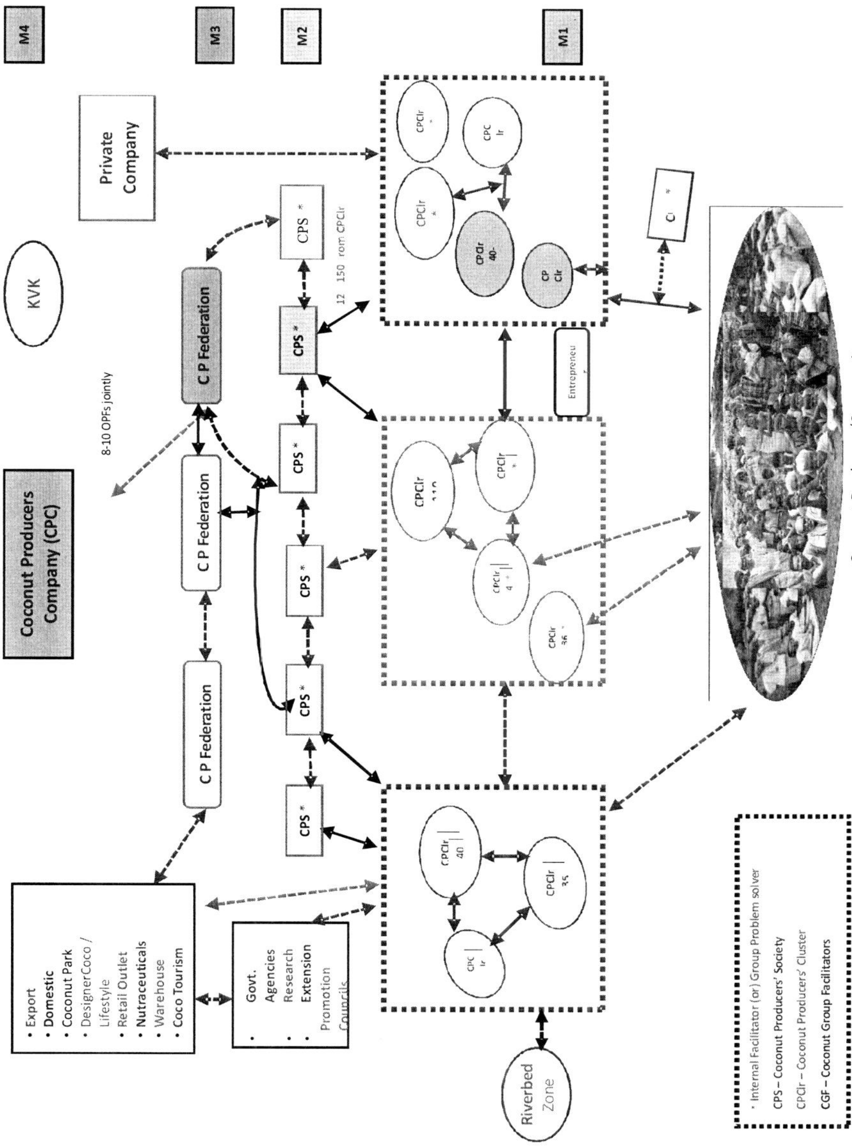

Fig. 3. Institutionalizing coconut sector: production to profitability

Moreover, the objective of meso-level, represents the formation of federation for collective information sharing, scale of production, processing and marketing. The final link of M1, M2 and M3 will lead to scale of marketing of value added products through M4 mega level to achieve scale of pricing for cluster. The critical component for sustaining cluster approach is nomination, nurturing and developing rural leader(s) for sustainable function of cluster approach. The leadership role in the cluster approach should enhance dynamism in mobilizing popular participation amongst the members of cluster/ community. Nurturing of leader in effective communication, extending the required financial & non-financial support, rural leadership skill and linking with multi-stakeholders role is essential for functioning of the cluster.

The interface and the involved actors between the micro, meso, macro and mega levels need to integrate the visioning, need identification, articulation and prioritisation of cluster. The bottom-up development network at the micro-level needs flexibility that shall be provided through multi-sectoral and pluralistic development within sector perspective once the formation of M4 cluster approach is developed and business value chain evolved across four levels. The FPO-M4 system may link with 'park' concept such as coconut-park as illustrative below in Fig.4. The concept of 'coconut park' will facilitate producers group, federation to directly link with coconut park for value addition and marketing, besides the facilities available with coconut producers' company.

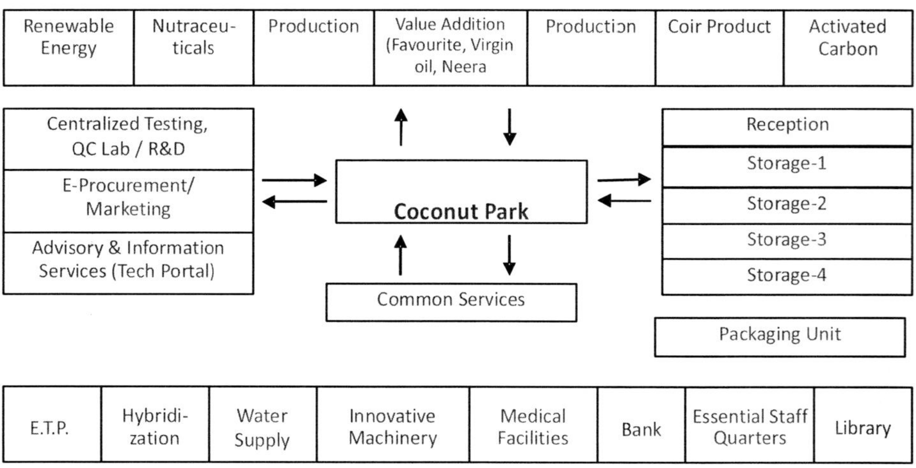

Fig. 4. Schematic profile of a coconut park

Conclusion

The producers' collective (e.g.: Coconut FPOs) should develop many opportunities to strengthen Commodity Business, improve the quality and relevance of services by drawing on holistic enterprises, association of farming community etc., for the improvement and performance of the sector. There are many combinations available to strengthen local action capabilities. Instead of relying primarily on inadequate members of relatively costly professionals, FPO led sub-professionals or para-professionals should advocate less formally trained and much less costly, local persons to extend advisory service to producers' collective.

The development organizations have already begin to support, and project that employ paraprofessional or indigenous non-civil service personnel to improve the cost effectiveness of paraprofessional service at producers' collective level. Paraprofessionals selected by, compensate by, accountable to producers' collective can increase the relevance and cost effectiveness of services while enhancing local self-reliance. Paraprofessional's role, function and skills as frontline service become increasingly the responsibility of Grassroots Competitive Advantage (GCA). The purpose of this section is to explore Strategic Rural Leadership programme (SRLP) for Karnataka rural citizens can increase the cost effectiveness of government expenditures in agri-commodity business sector through innovative grassroots staffing and organization patterns. These staffing innovation using local personnel or paraprofessional do involve a modest operating cost as compared with cost of normal civil service staff. The paraprofessionals are persons of indigenous, to the service area with the limited formal schooling and limited pre service training. The paraprofessional's works under the guidance and supervision of some professionals and academic institution such as IIPM. The sub professionals are men (including women) who are graduates from secondary school and 2-3 years of technical education in institutes or colleges. Normally they have full time and personable public functionaries and related persons. For e.g., sub professionals are rural leaders, progressive growers, retired schoolteachers, etc., who are in direct with the members of grassroots institution (FPOs). The important distinction is paraprofessionals are responsible and accountable to the neighbors, not to bureaucratic agency.

In a new emphasis on institution building or institution development, an attention is focused on strengthening producers' collective and its capacity building through paraprofessional (local action capability). The development offers detailed operational plan and establishment of a monetary system to check on the processes of the institution building and implementation required for the component of socio-environmental and cultural feasibility. The factors of producers' collective should be on sustainability of the local organizations, not on perpetuation of external support at all costs. Even if the meso-level organizations are not fully matured, still phasing out won't make a big difference as such support can be provided to them during their operational period. As it is a process and even the most mature (FPO) organizations have to constantly work on the capacity building of its staff and effectiveness of the systems. The most important factors, which influence the sustainability of producers' society/federation/company, are given below:

- Personal capacity building (skill in public speaking, community issues, interagency co-ordination)
- Popular participation and co-operation (ability to co-operate well with members).
- Commitment for innovativeness, skills related to fund raising and grassroots development process.
- Attention and skill in politics, selected issues and fund mobilization
- Knowledge on technical, leadership skills and organizational aspects
- Level of education, & IIPM-B led Karnataka Rural Leadership Program (KRLP) certification and
- Professional contacts for mobilizing aspects

Implications

The policy planners and development organizations cherish the myth that resource-limited people do not have the spirit to thrift, but recent reports from FPO/producers' collective of the globe challenge these aspects. Since FPO in India are informal groups, their legal status has not been defined. In the light of the above discussion, the prospects of FPO are manifold.

Firstly, FPO are self-sustained system of community (or) interest group (Producers' collective) and free from government bureaucracy. Secondly, these groups should be promoted by Village Based Organizations (VBO) such as association, producers' company/ federation, clubs etc., as its sub-group for cluster-based activities. Thirdly, the social cohesion in the group should give a sense of responsibility to its members as the "Best Rural Leaders" of the region. Fourthly, it should facilitate and enhance the **"process of empowerment for the functional knowledge"** to sustain its life cycle through participatory processes for initial period. Fifthly, the principle of self-selection by members in order to have social cohesiveness of the group is widely recognized. Sixthly, the sponsoring agency is expected to foster true (direct) democratic culture where all the members participate actively by taking part in the debate and decision making process which is possible only in FPO. Finally, FPO / Producers' collective promotion and nurturing are time consuming process.

Under the above situation, fixing un-realistic kind of target fixing in promotion of people participation goes against the basic principles of Producers' collective formation and deleterious impact on the sustainability of the People Processes for Development (PPD).

The producers' cluster beyond production should also develop many opportunities to strengthen commodity business to improve the quality, scale of production and marketing, certification, tourism, nutraceutical and other associated services. In a new emphasis on producers' cluster or institution building development, attention is given on strengthening the producers' cluster and its capacity building through (local action capability) rural leadership program to be initiated by IIPMB. The major role of clusters group should be on sustainability of the sector and not on perpetuation of external support at all costs. Even if the meso-level organizations are not fully matured, phasing out won't make a big difference, as such support can be provided to them during their operational period. As it is a process and even the most mature organizations have to constantly work on the capacity building of its members for effectiveness of the systems.

References

1. Bandhyopadhyay, D, Yugandhan, B.N and Mukherjee, A. 2002. Convergence of Programmes by Empowering SHGs and PRIs. *Economic and Political Weekly*. Pp.2556-2561.
2. Dhanakumar, V.G. 1994. Staff Decision Making Patterns, Village Leadership Performance, and Local Institutionalization Processes in Agricultural and Rural Development Programs. *The U.S. Journal of Agriculture, Food and Human Values Society*. 11(2 & 3). Pp. 67-74.
3. Dhanakumar, V.G. 1995. A Comparative Analysis of Institution Building for ARD in the Developing Countries. *Qty.Journal of International Agriculture*, Berlin.
4. Dhanakumar, V.G. 1999. An Assessment of Rubber Producers' Society (RPS): A Report on Grassroots Institution Building. Submitted to the World Bank and Rubber Board of India. Indian Institute of Plantation Management, Bangalore.
5. Dhanakumar, V.G. and Krishna Kumar A.K. 2000. An Assessment of Rubber Producers' Societies, (Self Help Groups of Small Growers) in the Rubber Plantation Sector in India. Proceedings of International Planters Conference, Malaysia. 17-20 May. Pp.491-502.
6. Dhanakumar, V.G. 2001. Grassroot Organization Developments & Management. *Management & Change*. 5(1). Institute for Integrated Learning in Management.
7. Dasgupta, R. 2001. Working and Impart of Rural Self-help groups and other forms of Micro Financing. Report of National Institute of Bank Management. Pp.371.
8. Esman, M.J.1980. Paraprofessionals and Rural Development. Cornell University. Centre for International Studies. Rural Development Committee.
9. Narayanswamy, N., Manivel.S and Bhaskar, B. 2003. Networking SHG's and Cooperatives. An Analysis of Strengths and Weaknesses. *Journal of Rural Development*.22(3). Pp.333-344.
10. White, W.F. 1981. Participatory Approach to Agricultural Development. Cornell University. Centre for International Studies Report.

❑❑❑

Chapter 13

Our Experience in Establishing and Making the FPOs Sustainable

Rajesh Urkude

mKRISHI, TCS

From operational perspective FPO has to be two/three tier organization

- SHGs/ Farmer Groups/ Farmer Clubs at village level
- Cluster of villages - Representative Director
- BoD, CEO and operational staff

Operational Area

There is cost to run an FPO, too small organizations are not viable.With experience of working in different states, it is felt that the average reach of an organization should be from 5 to 30 villages in cluster, with average size ranging from 1500 to 3000 farmers. We have observed that at any point of time one third of the members are actively participating with FPO. FPO should have different categories of membership; we encouraged primary membership latter on transforming to shareholders.Preferably the cluster should in single administrative unit i.e., block.

Identifying right directors and maintaining democratic set up

- Geographical representation
- Leadership qualities
- Organization before self-sacrifice: There are many farmers willing to dedicatedly contribute in organization building but do not want to accept the responsible positions, hence farmers with vested interest sneak into critical positions, many a times such

farmers are ready to pay premium for inducting them as director. Never fill positions of all 12 directors in first go, as the organization grows and spreads there has to be provision for onboarding new directors. From second term the directors should be first selected on specific criteria and then qualifying directors should be allowed to contest the election.

Role of POPI

For last 4-5 decades Most of these NGOs have been thriving on the grants from NABARD. NABARD has to move from fund / grand based approach to making even NGOs sustainable. In most of the government funded projects the capacity of the beneficiaries is supposed to be built, but we observe that most of the implementing agencies has built their capacities, now is the time these NGOs pass on their knowledge to FPOs and provide handholding in initial stages. We have to remember that FPO is all about social entrepreneurship development. They need to be supported by the organizations with entrepreneurial mindset. Now that we have given this task to NGOs, we need to sensitize them to this fact.It has been observed that most of the FPOs are not able to take independent decisions for two reasons, either they have not been taught how to take decisions or there is too much of interference from POPI.

FPO a democratic organization

In the initial establishment phase capital is scare so are other resources, including human resources. This is tough phase. Assistance in present format from various government agencies viz. SFAC/ World Bank funded projects or even NABARD are not enough.In the initial phase FPOs need to learn to live with limited resources, this brings in financial discipline but adequate financial and manpower support has to be seriously thought upon.In our experiments we had Project Manager (generally an agricultural graduate or PG) and couple of field executives (generally agriculture diploma holders), on the condition that they will have to earn their salaries from fourth year from FPO as they are seconded to FPO payroll.Usually the team engaged with 500, 1000 to 2000 farmers in first three years of their operations starting with 10, 15 to 25 villages. mKRISHI technology platform came very handy in engaging with these farmers. Initially we registered all willing farmers from villages as primary members (individually or gropes, if there was none we encouraged them to form one), in first six to eight months we intervention we introduced the FPO concept, simultaneously identifying farmers with leadership qualities for director positions. The regions where cooperative movement has flourished, the farmers have proactively accepted the concept. In many regions the concept is taking longer than expected because of local sociopolitical conditions, including caste system.We have seen many of the BoD slough it out along with mentor organizations, on the other side we are also seeing mushrooming growth of these FPOs only for syphoning off the government benefits and subsidies. FPO has become handy tool of these promoters (read politicians and there followers) to syphon off government schemes and subsidies. On the last count six months before Maharashtra had over 1050 registered FPOs.NABARD should come up with governance guidelines for FPOs on the lines of one suggested by Lodha Committee for The Board of Control for Cricket in India. Even the guidelines followed Mahi Dairy, Rajkot is worth looking at.

Revenue Generation Activities

Revenue generation activities is critical for sustenance of FPO, but it takes anywhere between 2-3 years to initiate first such activity. Most of the FPOs have started with input marketing but most of them have failed to make much headway for multiple reasons. Some of the challenges of the trade are advance booking, cash discounts, market volatility, etc. negligible margins in fertilizer business and farmer buying behavior on credit add to the woes. The trade distribution channel is ruthless, hence FPOs here too needs to adopt a collective bargaining strategy. Harvested produce is their main money spinner, hence FPOs should focus on this activity from day one. It takes substantial time to learn the tricks of the business and establish forward linkages with the market. While doing so most FPOs face working capital crunch, genuine buyers, etc. The FPOs who provided consistently quality produce has survived the competition from traders.

Last but not least value addition is the only viable and solution for FPOs so that member farmers benefitting from this exercise. This has amply been proven by success of dairy and sugar cooperatives. Hence it is imperative that FPOs migrate to value addition over a passage of time, this also helps them tide through glut situation by converting their raw material into values added products.

Other peripheral services to support revenue of FPOs

1. **Insurance:** (they need to be made corporate insurance agents, this will also help in penetrating insurance to non loanee farmers, which is major challenge
2. **Loan Agents / Business Correspondent:** They have captive user base, hence some experiments has to be done in this direction. FPOs should be encouraged to take up financial literacy in their working area for their member farmers
3. **Soil Testing Labs:** Many FPOs has purchased Mridaparikshak labs, if managed properly it's another revenue earning source for them. Government agencies should encourage soil testing by providing subsidy for doing the tests (RCF example)
4. **Forward Linkage:** Though farm produce is the biggest asset the FPOs has but we have to first understand how this channel works. It is easy to say that we will eliminate intermediaries but in practice it's easily said than done.
5. **Primary / Secondary Processing:** Singly or collectively FPOs has to achieve economies of scale in value addition viz. cotton ginning, dal mills, *Gud*, jams, pickles, puree, IQF, packing and retailing of perishables and nonperishable items through different channels, exports, etc.

In any system the intermediaries are existing because alternatives are not existing, the business systems have evolved and are flourishing because of dire need, in every sphere of life the law of survival of the fittest is equally applicable, hence FPOs has to enter into such spaces and occupy them. Some of the initiatives are financial lending, insurance, agro advisory consultancy, input supply and other village level related services though different models, grocery store, trading and value addition.

FPOs need to think differently. But it has been observed that unless these organizations enter into value addition they have little scope to scale up. Every FPO needs to study its ecosystem. It needs to look at the value chain. There are many commodities wherein FPOs can replace a trader, but they need financial support for their working capital needs. In few locations Tata Trusts has experimented with financial support. There has been mushrooming growth of micro financing agencies distributing credit at 22% interest rate and above, this has to be avoided and institutional credit linkage established.

In many cases the experiment failed because of syphoning of funds and non-ethical practices, many FPOs were not fast to respond to the dynamic markets. Some of them have started working closely with NCDEX but they have to operate consistently, its long way to go before we see success models. Few FPOs could do procurement under MSP, but its event based activity. We have connected much corn based FPOs to poultry feed manufacturers, the model is evolving but it comes with the dangers of usual market volatility. All this is part of learning; FPOs need to learn to trade. One of the ways is to become part of the system in small way, learn the tricks of the business and move up the value chain.

If we look at the success of diary and sugarcane value chain they are single product focused and very robust value chains, it is not the same with FPOs. Many of them are working in multiple commodities, the focus is missing. There is good scope for many to get into dal processing, cotton ginning, roller flourmills, etc. These all are capital-intensive projects, but if we want FPOs to sustain they need to get into value addition, they just cannot carry for too long a primary role

If the schemes under NCDC, MoFP for food park are doling out Crores of rupees to sugar and diary cooperatives/individual entrepreneurs/corporates then why not to FPOs.

Licenses

To conduct different types of businesses the FPOs has to procure different licenses; based on their experience many state governments are enabling these activities. SFAC, Government of Maharashtra Government of Madhya Pradesh has been in the forefront of establishing FPOs through their flagship MACP project. Other state governments and agencies have to learn from their experience.

Agriculture being state subject many of them has different rules and regulations. These rules and regulations need to be streamlined as has been done for GST. In couple of our experiments we observed that FPOs could not do business even after having Direct Marketing license. Agriculture Produce Market Committee (APMC) is central to every farmer to conduct business so is it for FPOs. Mostly farmer depends on APMCs for selling his produce. In couple of cases FPOS were not able to procure agricultural produce from APMC to be supplied to processors, as they are not the members of APMC. FPOs should be helped in getting APMC license so that they can conduct business in these mandi's at par with traders. Mandi's are under control of local body, the traders to safeguard their wrested interest do not allow FPOs to operate in APMC premises leading to corrupt practices.

Subsidy Disbursement and Exemptions

RKVY funded PPP Aid projects has assigned chunk of funds to boost value addition in agriculture and allied sector. It should not be prerogative of corporates but FPOs should also be allowed to participate and innovative models should be encouraged.In many a cases Subsidy is disbursed on completion of project which helps in effective and actual use of subsidy. It is to be noted that many projects undertaken by FPOs are capital intensive hence practice of phase wise disbursement and monitoring should be adopted instead.

State level FPC

State level FPCs in Maharashtra and MP are functioning. It is interesting to study their progress and achievements since there establishment. It needs to be further studied and discussed how their functioning could made transparent and effective. Has these organizations fulfilled the aim with which they were formed. What type of leadership is manning these organizations and how they could be made more effective?

Training and capacity building

If we look at the state of Maharashtra, steadily enough expertise has been built at district level. This was possible because of World Bank funded MACP project. NABARD too established 6 centers to cater to new FPOs being established in different parts of the country. There is quite a lot learning from these experiments. Now that there are enough new organizations being set up at district and regional level such training and capacity building activities should be conducted at local level rather than spending huge amounts on residential trainings. Major costs are incurred towards travel and accommodation. These funds can be best utilized in other ways. Most of the FPOs has asked for on job training rather than class room trainings. In first phase many of these training centers themselves did not have right set of manpower and resources to handle trainings of CEOs and BoD.

State Agricultural Universities at regional level and national institutions viz., MANAGE, IRMA, National Institute of Rural Development and Panchayati Raj, IIMA, IIML BIRD should be promoted at national level.

Exposure visits

Exposure visits to successful FPOs plays very important role in motivating aspirants and new organizations, hence such initiatives should not only be encouraged but organized thoughtfully with clear mandate, objectives and clear out comes rather than making it annual rituals and picnics. More corporate destinations, food parks, processing facilities, marketing networks, national mandi's, should be included as places to visit. The apex organization like NABARD or institutes mentioned above should come up with list such locations, clusters, route maps, etc. These visiting locations should be incentivized for entertaining FPOs.

Chapter 14

Land Resources Information for Enhancing Efficiency and Sustainability of FPOs in Karnataka

Rajendra Hegde

NBSS & LUP, Bengaluru

The Department of Horticulture, Government of Karnataka has initiated formation of FPO during the year 2014-15 with the support of FPO nodal agency Small Farmers Agribusiness Consortium (SFAC), New Delhi by mobilizing the Farmer Clusters already formed by the Department and other farmers into the larger groups of 1000 members to form a Farmer Producer Company under Companies Act 2013(Ref). In Karnataka presently, 75 FPOs are registered under Companies Act 2013 and remaining 17 FPOs are at different stages of registration. These FPOs have covered about 70,000 farmers majority of which are small and marginal farmers and economically backward class. The resource institute, FPO chief executive officers and board of directors conduct regular meetings at Farmers Interest Groups level to discuss on the challenges faced by them. Suggestions are offered on integrated pest and nutrient management practices, farm mechanization etc., for adoption of sustainable practices and money-saving solutions. The idea was to get farmers act collectively and obtain group discounts on the purchase of horticultural inputs and, thus, eventually get a better price for the sale of their produce. FPOs have enabled farmers to exercise more control over their input requirement and output marketing. Department of Horticulture with the help of Resources Institutions facilitated seed, fertilizer, pesticide and Agricultural Produce Market Committee trade licenses. Thus, FPOs have established backward input linkages by stocking the entire range of inputs—including seeds, fertilizers, bio-pesticides, pesticides and farm equipments for its members. Both the member and non-member farmers have received numerous benefits from this intervention which included easy accessibility to quality agricultural inputs without the fear of spurious products and assured availability of the inputs throughout the season. In addition, it ensured affordable and fair prices that led to farmers saving 5-10% on the input costs alone, as well as reduced the amount of time spent to get these inputs, since they are now available in the village itself. Savings in input and labour cost reduces the cost of production by nearly 20%, thereby the farmers can get more profit to the produce they grow. After realizing the need to support farmers after FPO registration, the Department of Horticulture during 2015-16 started facilitating FPOs by giving revolving fund of Rs. 15.00 lakhs and 90 per cent subsidy support (maximum of Rs. 22.50 lakhs) to establish custom hiring centers (CHC). 67 FPOs have received revolving fund and 59 FPO have been supported to establish CHCs (Ref 2).

Through FPO, farmers are collectivized, sharing their farming experiences with each other. Production cost has been lowered due to low cost inputs available in FPO where they can access easily. Encouraging establishment of custom hiring centers by FPOs which offer machinery at affordable rates has improved the productivity and reduced the drudgery of farm labour. CHC provides access to costly farm machinery for small and marginal farmers as well as facilitates timeliness in farm operations and efficient use of inputs. Overall the production cost has been reduced, which in turn adds on to the profit (Ref).

Land resources based improved land management practices are key for the yield improvement and economic profit. However a single set of recommendations were not be appropriate for the different resource endowments farmers. Land resources information can be used to predict or estimate the potentials and limitations of soils for many specific uses. A soil survey includes an important part of the information that is used to make workable plans for land management. The information must be interpreted to be useable by professional planners and others. A knowledge of the characteristics of the soil helps in determining the crop choices, irrigation scheduling, soil amendment needs, general drainage requirements, and crop management practices needed to maintain optimal soil conditions for plant growth. Soil surveys are also helpful for locating possible sources of sand, gravel, or topsoil. Soil Survey information is an important component of technology transfer.

This paper concerns identifying the land resources constraints and potentials status Nanjevanapura village in Udigala Horticultural Producers Company, Chamarajnagar district, Katrnataka was established in 2016 by the Department of Horticulture, Government of Karnataka. This company program was formed to ensure the sustainable development of horticulture sectors. Keeping in view the objectives of FPO, an attempt was made (1) To study land resources inventory and (2) To provide the land resources information for enhancing the efficiency in horticultural production.

Methodology

Nanjedevanapura village is one of the villages in Udigala Horticultural Producer Organization, in Chamarajanagar Taluk and district in Karnataka. The study area is located in the southern part of Karnataka in Chamarajanagar, taluk in Chamarajanagara district, Karnataka, it covers an area of about 808 ha. The area forms part of granite and granite gneiss landscape. The climate is semiarid and categorised as drought-prone with average annual rainfall of 754 mm. Generally, the length of crop growing period (LGP) is 150 days.

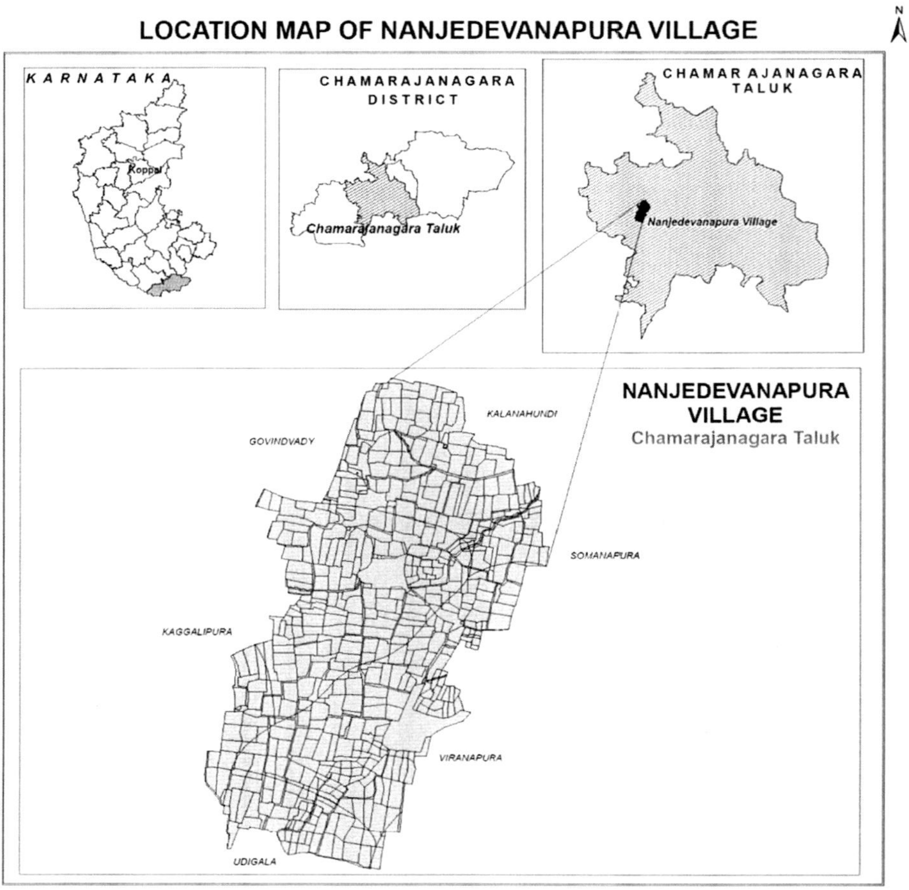

Fig. 1. Location map of Nanjevevanapura Village.

Mapping of Land Resources

Land Resources Inventory for site-specific planning and development of watersheds is being carried out by using digitised cadastral map and False Colour Composites (FCC) of Cartosat-1 and LISS-IV merged satellite data. The FCCs were visually interpreted using image interpretation elements along with all the collateral data available for the area.

The delineated physiographic boundaries were transferred on to a cadastral map overlaid on satellite imagery and used as base map for mapping soils. Intensive traversing of each physiographic unit like hills, uplands and low lands was carried out. Based on the soil variability observed on the surface, transects were selected across the slope covering all the physiographic units identified in the watershed. In the selected transect, soil profiles were located at closely spaced intervals to take care of any change in the land features like break in slope, erosion, gravel, stones, saline/alkali etc. In the selected sites, soil profiles (vertical cut showing the soil layers from surface to the rock/water) were dug up to 200 cm or to the depth limited by rock or hard substratum. The profiles were studied and described in detail for all their morphological and physical characteristics and also the soil and site characteristics. Apart from the transect study, soil profiles were also studied at random, almost like in a grid pattern outside the transect areas to validate the soil map unit boundaries. Based on soil characteristics, the soils were classified upto series level and grouped into different soil series. Soil series is the most homogeneous unit having similar horizons and properties, and behaves similarly for a given level of management. Soil depth, texture, colour, kind of soil horizon and its sequence, gravel, stones, calcareousness, nature of substratum etc. were used as differentiating characteristics for identifying soil series.

The soils were mapped as phases of soil series. The area under each soil series was further separated into soil phases (management units) by traversing each and every land parcel in all directions in the watersheds and their boundaries delineated on the cadastral map based on the variations observed in the texture of the surface soil, presence of gravel/stones, slope, erosion etc. A soil phase is a subdivision of soil series based mostly on surface soil features that affect its use and management. The soil mapping units are shown on the map in the form of symbols. The soil map shows the spatial distribution and their area extent of different soil mapping units (soil phases) identified under each soil series. All the land parcels/survey numbers included in one soil phase will have similar soil and site characteristics that require same management and respond similarly for a given level of management. Soil samples were collected from representative master profiles for each soil series for laboratory characterisation of physical and chemical properties. Surface soil samples collected from farmers fields for fertility status (macro and micronutrients) at 250 m (one sample for every 6.25 ha area) grid interval were analysed for different nutrient elements.

Table 1. Land Resources Information for Production Planning

Land resources Status	Soil Conservation / Management	Convergence with line departments
Type and Extent of soil distributions	Improved land management practices (e.g., improved agronomic practices, and various soil and water conservation measures) can improve soil fertility and structure, adding high amounts of biomass to the soil, causing minimal soil disturbance, conserving soil and water, enhancing activity and diversity of soil fauna, and strengthening mechanisms of elemental cycling.	Karnataka Watershed Department can undertake the implementation of soil and water conservation measures
Soil fertility status	Integrated nutrient management (e.g. precision farming including efficient fertilizer application based on crop and site specific nutrient balance analysis, split application, timing). Proper management of organic soils (e.g. avoiding deep drainage and deep ploughing, row crops and tubers and maintaining a shallower water table) can reduce N2 O and CH4 emissions.	Department of horticulture for Providing necessary soil nutrients.

Results

Significant findings

The analysis and interpretation of LRI data helped in identifying the potential and problem areas in the project villages

Potential areas

- About 28% area has deep to very deep (>150 cm depth) soils (Fig. 1).
- About 49% area has non-gravelly (<15% gravel) soils.
- About 80% area is very gently (1-3% slope) sloping lands.
- 88 % area is slightly eroded.

- About 49% area is high 0.75%) in soil organic carbon (Fig. 2).
- About 85 % area is medium to high (>145-337 kg/ha) in available potassium.
- Entire area is sufficient in manganese, copper and 19 % area is sufficient in zinc.

Problem areas

- About 61% of area is low (<0.6 ppm) in available sulphur. Such areas need to be applied with magnesium sulphate or gypsum or factamphos (P) fertiliser (13% sulphur) for 2-3 years for the deficiency to be corrected (Fig. 3).
- About 71% of area is low (<0.6 ppm) in available zinc. Such areas need to be applied with zinc sulphate for 2-3 years for the deficiency to be corrected (Fig. 4).
- About 51% of area is low (<4.5 ppm) in available iron. Such areas need to be applied with ferrous sulphate for 2-3 years for the deficiency to be corrected (Fig. 5).
- About 67% of area is low (<0.5 ppm) in available boron. Such areas need to be applied with borax for the plants to deficiency to be corrected (Fig. 6).

The Land Resource Information data thus helps in identifying the most potential and problem areas and if proper corrective measures are adopted, the productivity in the micro watershed can be enhanced on sustainable basis.

The Survey number wise soil fertility status is given for the Nangedevanapura village (Table 2). This information can help for estimating the fertilizer requirement and providing integrated soil nutrient management in crop production. The soil fertility information helps in planning the fertilizer demands for each farm households. An examples is worked out to show that how to use the soil fertility data for the FPO registered farmers of Nanjevavanapura village is shown in (Table 3).

Table 2. Survey number wise soil fertility status in Nanjedevanapura Village

S.No	Survey No	pH	OC	P_2O_5	K_2O	S	B	Cu	Fe	Mn	Zn
1	2	SA	M	L	M	L	L	S	S	S	D
2	3	SA	H	M	M	L	M	S	S	S	D
3	8	SAk	H	H	H	L	M	S	D	S	S
4	10	MAk	H	H	H	H	M	S	D	S	D
5	13	MAk	L	L	H	L	L	S	D	S	D
6	21	MAk	H	M	H	L	M	S	D	S	D
7	24	MAk	H	L	H	L	H	S	D	S	D
8	31	MAk	M	L	H	L	L	S	D	S	D
9	36	MAk	H	L	H	L	M	S	D	S	D
10	48	N	H	L	M	M	L	S	S	S	D
11	50	MAk	H	L	H	L	H	S	D	S	D
12	67	MAk	H	L	H	L	M	S	D	S	D
13	71	N	H	L	M	L	L	S	D	S	D
14	74	SAk	L	M	H	L	L	S	D	S	D
15	91	MAk	L	M	H	L	M	S	D	S	D
16	92	MAk	L	H	H	H	L	S	D	S	S
17	96	MAk	L	M	H	H	M	S	D	S	D
18	98	MAk	H	L	H	M	M	S	D	S	S
19	102	MAk	L	M	H	L	M	S	D	S	D
20	105	SAk	L	M	H	L	L	S	D	S	S
21	112	SAk	L	M	M	L	M	S	D	S	D
22	118	MAk	H	M	H	L	M	S	D	S	S
23	121	MAk	L	L	M	L	M	S	D	S	D
24	130	MAk	M	M	H	H	M	S	D	S	S
25	135	SA	H	L	L	L	L	S	S	S	D
26	138	N	H	L	M	L	L	S	D	S	D
27	143	N	H	L	M	L	L	S	D	S	D
28	149	SAk	L	M	H	L	L	S	D	S	D
29	151	SAk	H	M	H	L	L	S	D	S	S
30	153	MAk	H	L	M	M	L	S	D	S	D
31	159	MAk	H	L	H	L	L	S	S	S	D
32	167	N	L	M	L	L	L	S	S	S	S
33	169	SAk	H	L	M	L	L	S	D	S	D
34	171	MAk	M	L	M	L	L	S	D	S	D
35	173	SAk	H	H	H	M	L	S	D	S	S
36	178	N	L	L	L	L	L	S	S	S	D
37	179	MAk	L	L	L	L	L	S	D	S	D
38	186	MAk	H	M	H	M	L	S	D	S	D
39	194	MAk	L	L	M	M	L	S	D	S	D
40	196	MAk	H	L	H	L	L	S	D	S	D

41	202	MAk	H	H	H	L	L	S	D	S	D
42	205	MAk	H	H	H	L	L	S	D	S	S
43	216	MAk	H	L	H	L	M	S	S	S	D
44	219	MAk	L	L	M	L	L	S	D	S	D
45	225	MAk	H	L	M	L	L	S	D	S	D
46	229	MAk	H	M	H	L	L	S	D	S	D
47	232	SAk	H	L	H	L	M	S	D	S	S
48	246	MAk	H	L	M	L	L	S	D	S	D
49	248	MA	L	L	M	L	L	S	S	S	D
50	252	MAk	H	M	M	L	L	S	D	S	D
51	260	MAk	L	L	M	L	L	S	D	S	D
52	262	MAk	H	M	M	M	L	S	D	S	S
53	266	N	H	L	L	L	L	S	S	S	D
54	273	MAk	M	L	M	L	L	S	D	S	D
55	277	MAk	H	H	H	L	L	S	D	S	S
56	282	SA	H	L	L	L	L	S	S	S	D
57	284	MAk	L	L	M	L	L	S	D	S	D
58	288	SAk	H	L	M	L	L	S	D	S	D
59	291	MAk	H	M	H	L	L	S	D	S	D
60	296	SAk	M	L	L	L	L	S	S	S	D
61	302	MAk	H	L	H	M	L	S	D	S	S
62	304	MAk	H	L	M	L	L	S	D	S	D
63	308	MAk	H	H	H	M	L	S	D	S	S
64	311	MAk	L	M	H	L	L	S	D	S	D
65	312	N	H	L	M	L	L	S	S	S	D
66	315	MAk	H	L	H	L	L	S	S	S	D
67	317	N	M	L	L	M	L	S	S	S	D
68	319	SAk	H	L	M	L	L	S	S	S	D
69	324	SAk	H	M	H	L	L	S	D	S	S
70	328	N	M	L	L	M	L	S	S	S	D
71	331	N	H	L	M	L	L	S	S	S	D
72	332	N	H	L	M	L	L	S	S	S	D
73	334	N	M	L	L	M	L	S	S	S	D
74	340	MAk	H	M	H	M	L	S	D	S	D
75	346	MAk	L	L	H	M	L	S	S	S	D
76	353	MAk	H	L	H	H	M	S	D	S	D
77	362	MAk	L	M	M	M	L	S	D	S	D
78	365	SA	H	L	M	L	L	S	S	S	D
79	367	MAk	L	M	M	M	L	S	D	S	D
80	382	SA	H	M	M	L	M	S	S	S	D
81	387	SA	H	M	M	L	M	S	S	S	D
82	389	SAk	L	H	H	L	M	S	D	S	S

Table 3. Soil fertility data for the FPO registered farmers of Nanjevavanapura village

S.No.	Survey No	pH	OC	P_2O_5	K_2O	S	B	Cu	Fe	Mn	Zn
83	392	MAk	L	L	M	L	H	S	S	S	D
84	393	SA	H	M	M	L	M	S	S	S	D
85	397	MAk	L	M	H	L	M	S	D	S	D
86	400	MAk	M	M	H	M	M	S	D	S	S
87	401	MAk	M	L	M	H	L	S	D	S	D
88	404	SAk	L	L	M	L	L	S	D	S	D
89	406	N	H	L	M	L	M	S	D	S	D
90	415	N	M	L	M	L	L	S	S	S	D
91	417	MA	M	L	M	L	L	S	S	S	D
92	420	N	M	L	M	L	L	S	S	S	D
93	421	SA	M	L	L	L	L	S	S	S	D
94	426	SA	H	H	H	L	L	S	S	S	S
95	434	N	H	L	L	H	L	S	S	S	D
96	435	N	H	L	M	M	L	S	S	S	D
97	440	MAk	M	L	H	L	L	S	D	S	D
98	442	SA	M	L	M	M	L	S	S	S	D
99	444	MA	H	L	M	M	L	S	S	S	D
100	445	SAk	M	L	M	L	L	S	D	S	D
101	449	SA	M	L	M	M	L	S	D	S	D
102	454	SA	H	L	M	L	L	S	S	S	D
103	464	MA	M	L	L	M	L	S	S	S	D
104	467	SA	M	L	M	M	L	S	S	S	D
105	469	SA	L	L	M	M	L	S	S	S	D
106	473	N	H	L	H	L	L	S	S	S	D
107	477	MA	M	L	M	M	L	D	S	S	D
108	477	MA	M	L	M	L	L	S	S	S	D
109	478	N	H	L	M	M	L	S	D	S	D
110	485	MA	H	L	M	H	L	S	S	S	D
111	489	MAk	M	L	H	L	M	S	D	S	D
112	490	SAk	H	L	H	L	L	S	D	S	S
113	494	MAk	H	H	H	L	M	S	D	S	S
114	494	MAk	H	M	H	M	L	S	D	S	D
115	499	SA	M	L	M	M	L	S	S	S	D
116	501	SAk	M	M	H	L	M	S	D	S	D
117	520	N	M	L	L	M	L	S	D	S	D

Note: MA: Moderately acid (pH 5.5-6.0), MAk: Moderately alkaline (pH 7.8-8.4), N: Neutral (pH 6.5 -7.3), SA: Strongly acid (pH 5.0–5.5), SAk : Slightly alkaline (pH 7.3-7.8), L: Low, M: Medium, H:High, S: Suffient, D: Deficient.

In the absence of soil health status the farmers were followed the blanket recommendation based on advise of fertilizers dealer. The amount of fertilizer recommendation in tomato cultivation would resulted uniform application of costly soil nutrients (Table 4).

Table 4. Blanket recommendation for tomato cultivation

Name of the farmers	Survey No	Fertilizer requirements (Kg/acre)			Total Cost (Rs/ha)
		DAP	Urea	MOP	
Mallesh	2	217	132	167	7410
Madappa	8	217	132	167	7410
Shashidar	74	217	132	167	7410
Nagappa	92	217	132	167	7410
Basavanna	98	217	132	167	7410
K L Shanthappa	112	217	132	167	7410
Basavanayka	130	217	132	167	7410
Mahadevappa	485	217	132	167	7410

In the event of following the Land Resources Information for guiding fertilizers application in tomato cultivation the farmer wise fertilizer soil fertility status (Table 5) and fertilizer requirement for tomato cultivation can suggest site specific requirements (Table 6).

Table 5. Farmers wise soil fertility status in Nanjedevanapura village

Name of the farmers	Survey No	pH	OC	P_2O_5	K_2O	S	B	Cu	Fe	Mn	Zn
Mallesh	2	SA	M	L	M	L	L	S	S	S	D
Madappa	8	SAk	H	H	H	L	M	S	D	S	S
Shashidar	74	SAk	L	M	H	L	L	S	D	S	D
Nagappa	92	MAk	L	H	H	H	L	S	D	S	S
Basavanna	98	MAk	H	L	H	M	M	S	D	S	S
K L Shanthappa	112	SAk	L	M	M	L	M	S	D	S	D
Basavanayka	130	MAk	M	M	H	H	M	S	D	S	S
Mahadevappa	485	MA	H	L	M	H	L	S	S	S	D

Source: List of Farmers from CEO, Udigala FPO, Chamarajanagar

Table 6. Soil test based fertilizer recommendation for Tomato cultivation

Name of the farmers	Survey No	Soil Nutrients Requirement (Kg/acre)			Fertiliser requirement (Kg/acre)			Total Cost (Rs/ acre)
		N	P_2O_5	K_2O	DAP	Urea	MOP	
Mallesh	2	100.0	125.0	100.0	217	132	208	7868
Madappa	8	75.0	75.0	75.0	163	99	125	5557
Shashidar	74	125.0	100.0	75.0	272	165	167	8804
Nagappa	92	125.0	75.0	75.0	272	165	125	8346
Basavanna	98	75.0	125.0	75.0	163	99	208	6474
K L Shanthappa	112	125.0	100.0	100.0	272	165	167	8804
Basavanayka	130	100.0	100.0	75.0	217	132	167	7410
Mahadevappa	485	75.0	125.0	100.0	163	99	208	6474

The impact of change in fertilizer use and the cost of fertilizer saving are given in Table 7. This conclusively support in terms of avoiding cost of fertilizer misapplication and its impact on soil fertility degradation. It shows there by using Land resources information is possible in reducing the cost of cultivation, enhancing tomato productivity and long term sustainability of Farmers Producer Companies.

Table 7. Impact of Soil Test Based Fertilizer Recommendations in Tomato cultivation

Name of the farmers	Survey No	Soil Nutrients Misapplication(Kg/acre)			Total Cost of misapplication(Rs/acre)	Impact of misapplication on soil health
		DAP	Urea	MOP		
Mallesh	2	0.00	0.00	-41.67	-458	Deficit application of K
Madappa	8	54.35	33.08	41.67	1853	Excess application of NPK
Shashidar	74	-54.35	-33.08	0.00	-1394	Deficit of N and Excess of K
Nagappa	92	-54.35	-33.08	41.67	-936	Deficit of N and P and Excess of K
Basavanna	98	54.35	33.08	-41.67	936	Excess of N and P and Deficit of K
K L Shanthappa	112	-54.35	-33.08	0.00	-1394	Excess of N and P
Basavanayka	130	0.00	0.00	0.00	0	Optimum
Mahadevappa	485	54.35	33.08	-41.67	936	Excess of N and P and deficit of K

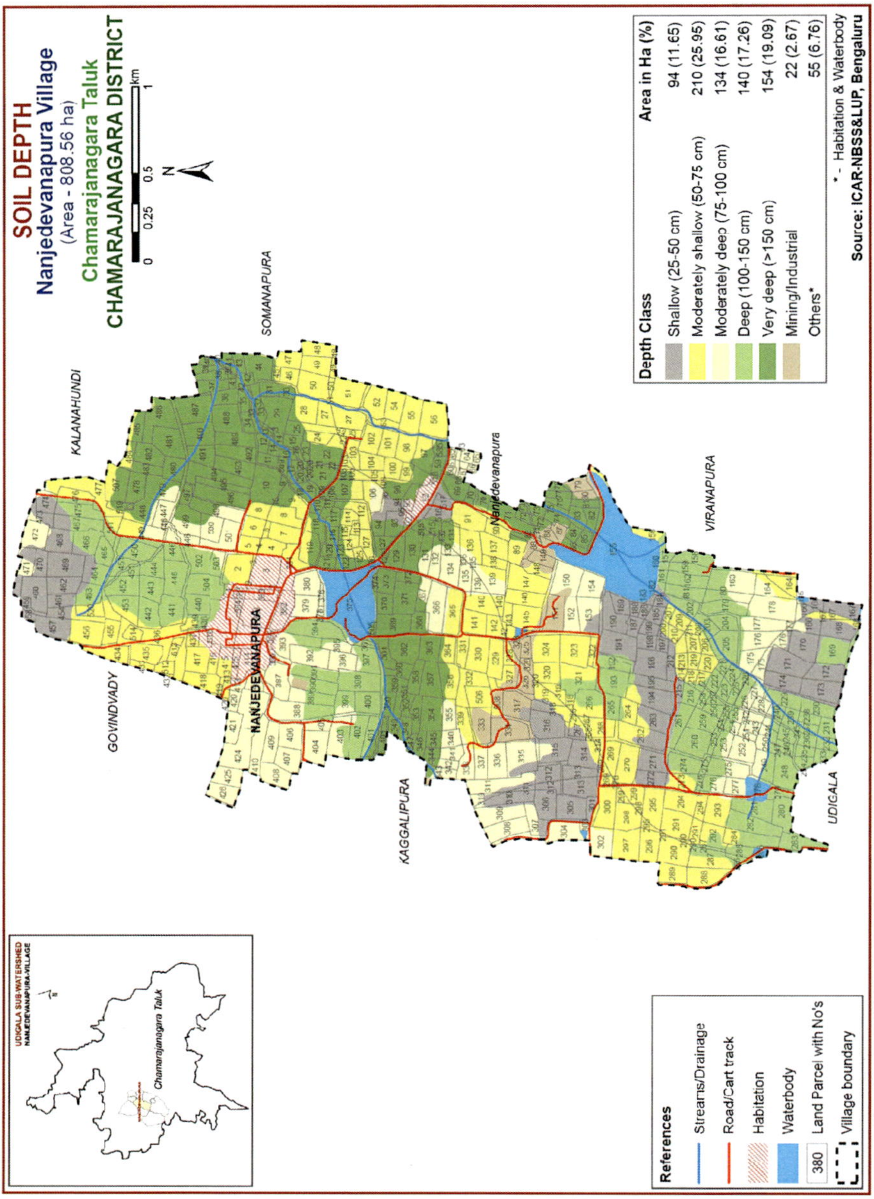

Fig. 2. Soil depth

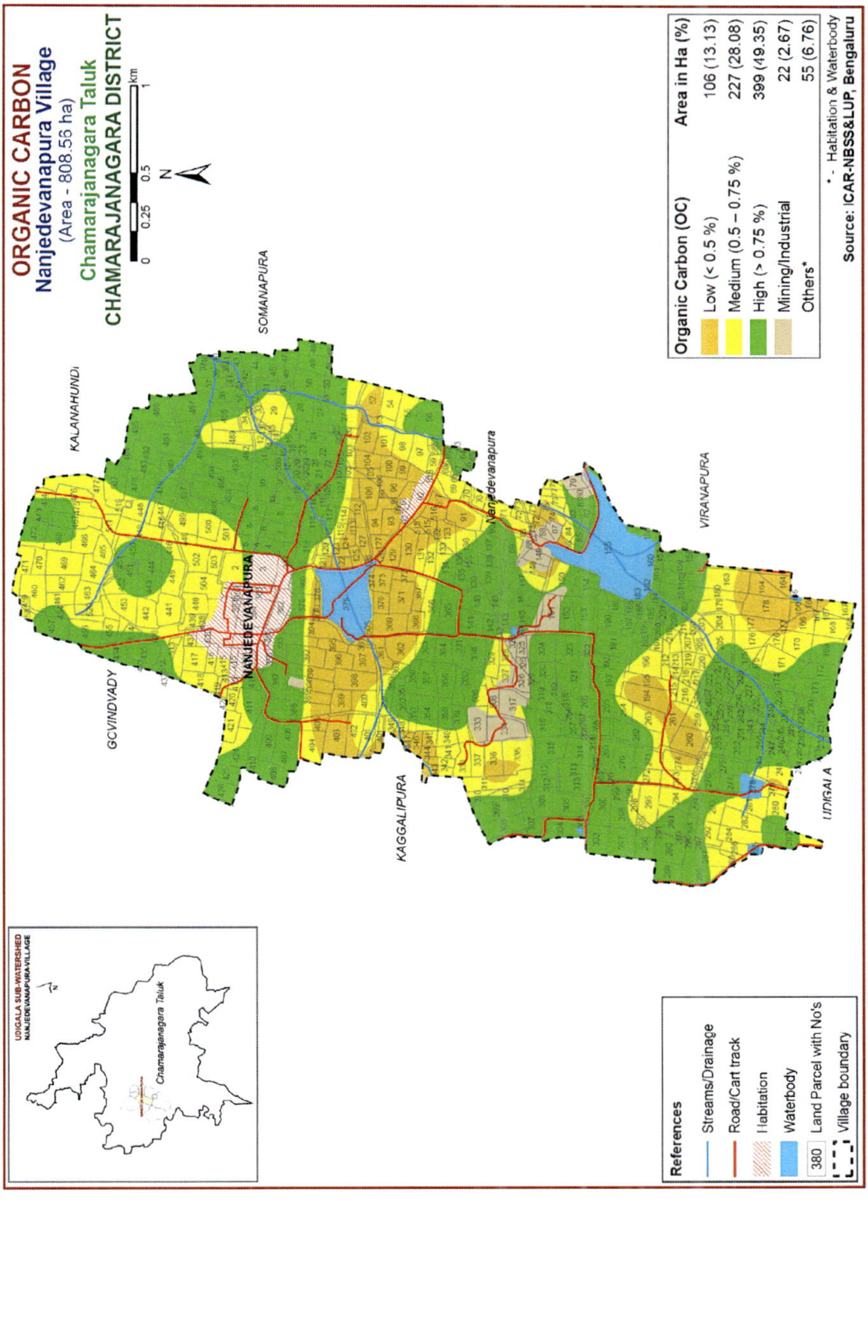

Fig. 3. Soil Organic Carbon

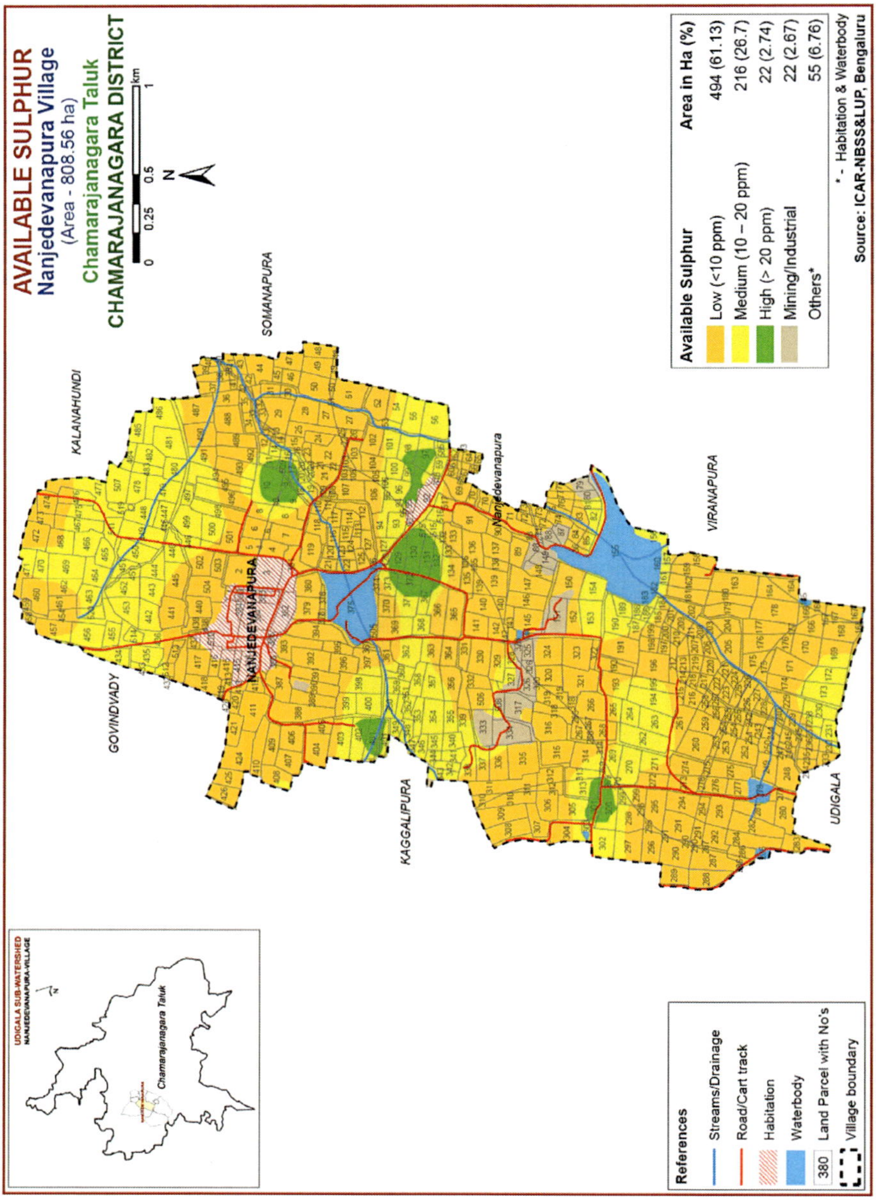

Fig. 4. Available Sulphur

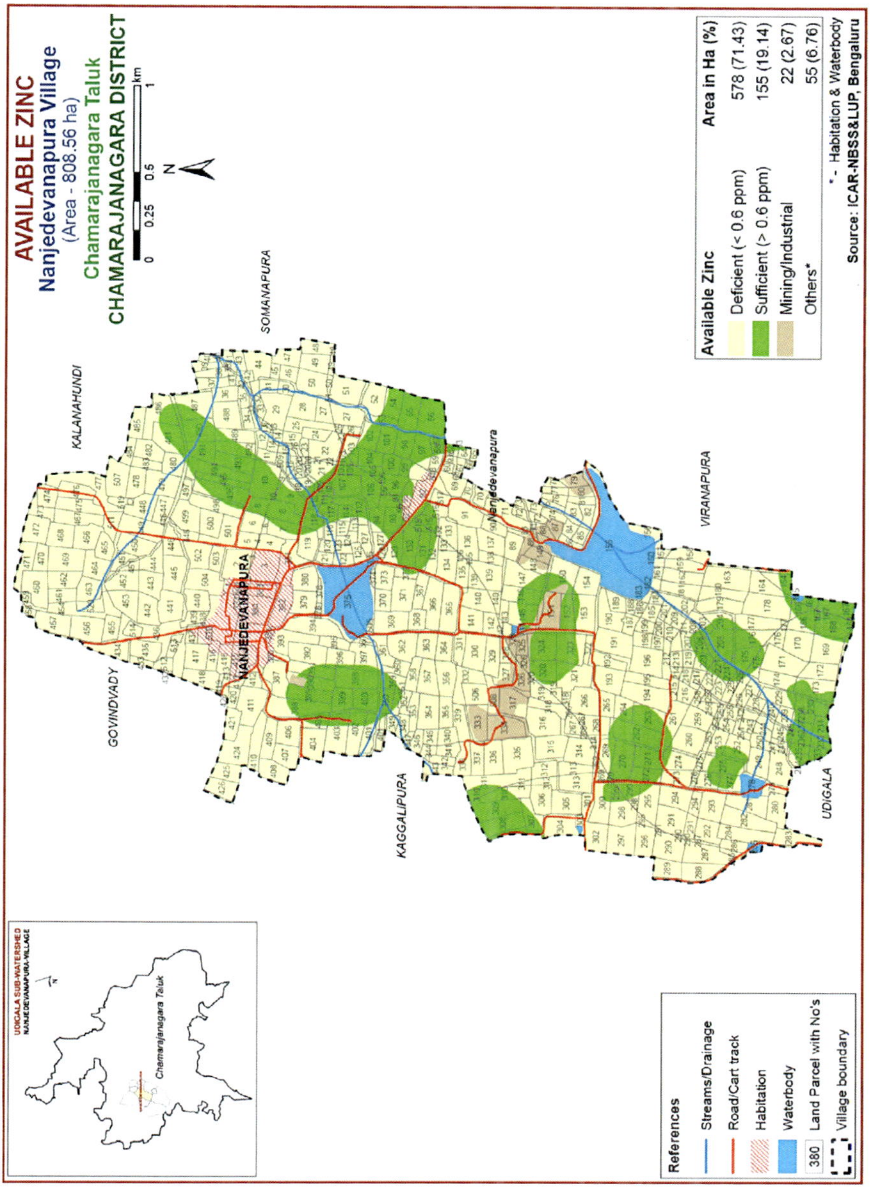

Fig. 5. Available Zinc

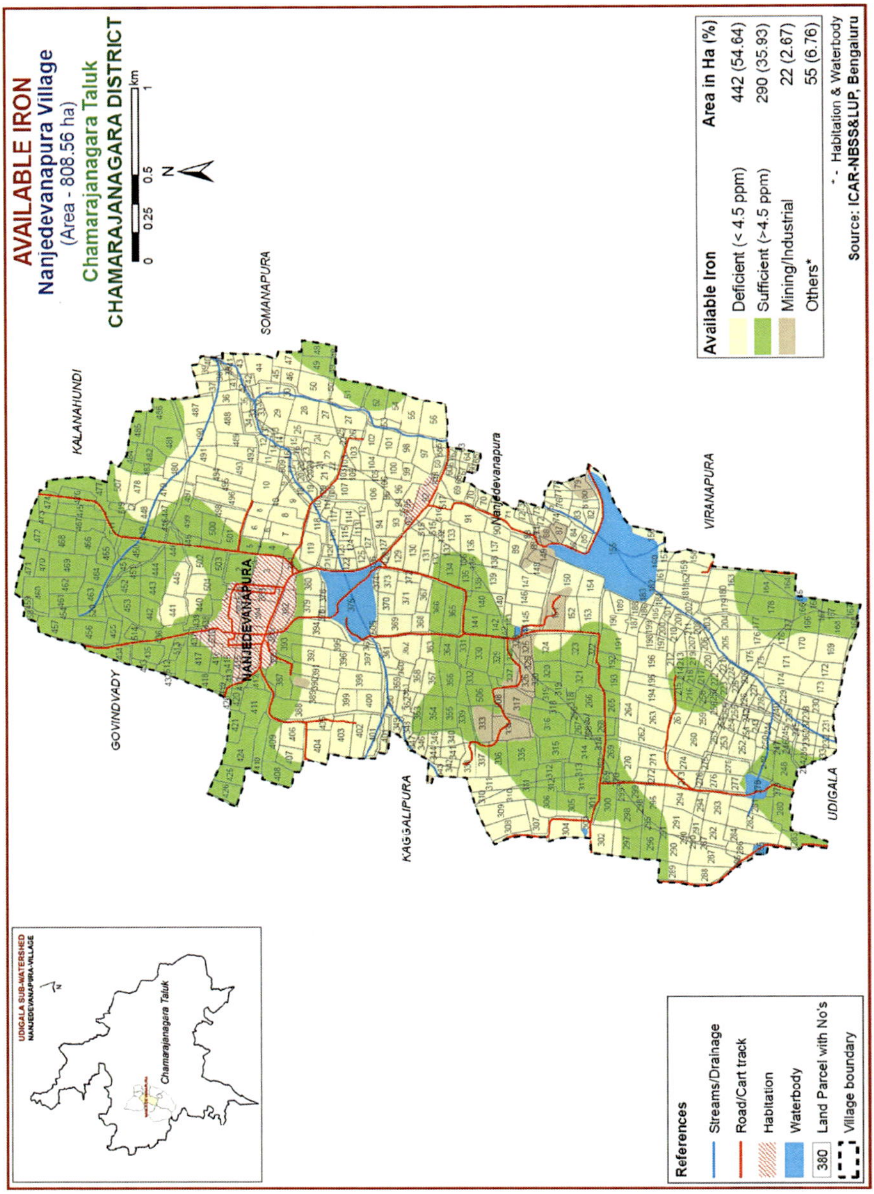

Fig. 6. Available Iron

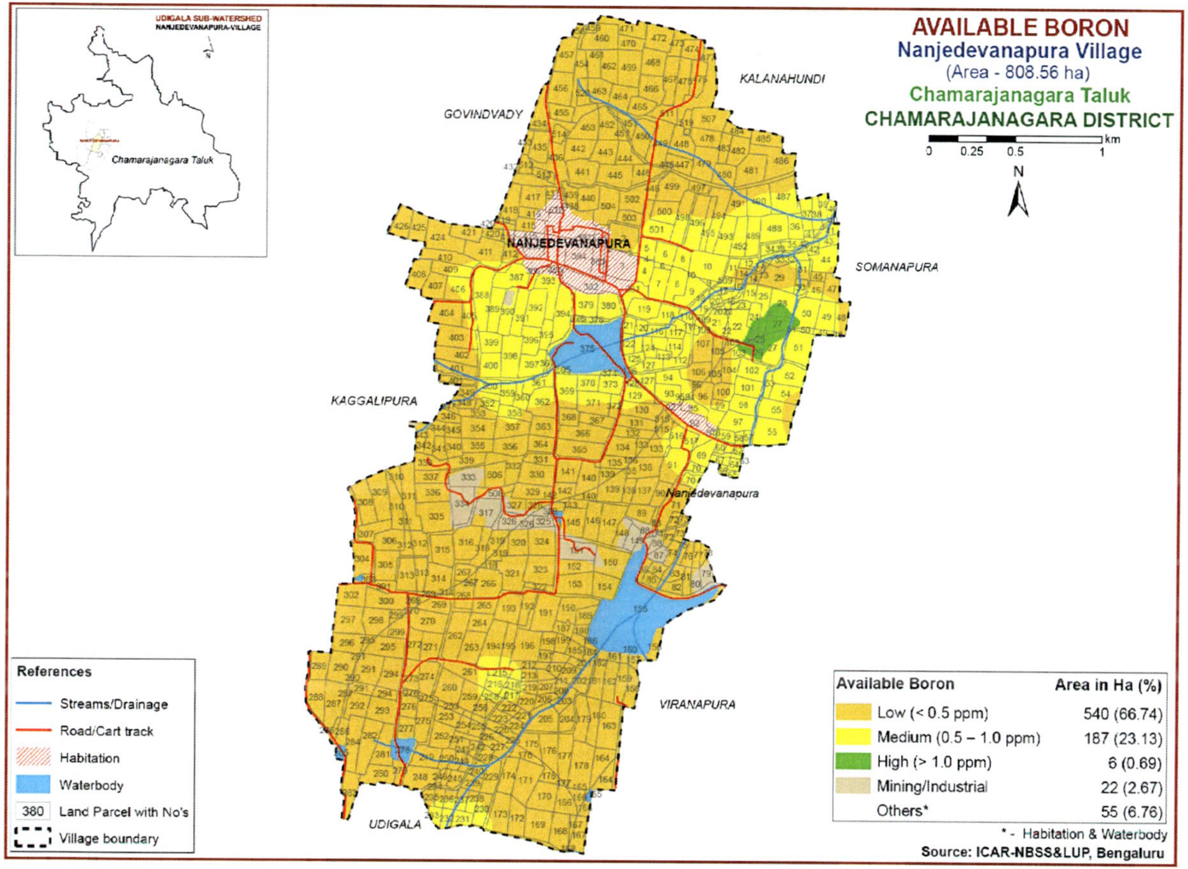

Fig. 7. Available Boron

References

1. http://sfacindia.com/PDFs/List-of-FPO%20identified-by-SFAC/List%20of%20FPOs%20in%20the%20State%20of%20Karnataka.pdf
2. http://rkvy.nic.in/(S(nd0uzghpmljf1qsqwlicizvp))/Uploads/SucessStory/KARNATAKA/2017/2017012054FPO%20Success%20Story%20(RKVY)final%20new.pdf
3. https://ccsinternship.files.wordpress.com/2013/05/293_contract-farming-through-farmer-producer-organizations_varun.pdf
4. http://www.nabard.org/auth/writereaddata/File/FARMER%20Producer%20Organisations.pdf

❑❑❑

Chapter 15

Convergence Approach through FPOs for Sustainability of Small Farmers in Andhra Pradesh

Subba Rao D V

NABARD Chair, ANGRAU, Guntur

The idea of Farmers Producers Organisation, FPO is a novel and ambitious initiative with long run vision. The sustainability criterion of FPO is important to provide continued benefits to the members. The initiative enables to adapt climate resilient models for sustainability of farmers, particularly small and marginal. It can handle the markets and act as a substitute to the existing middlemen at different levels in due course of time. To replace middlemen, all the functions performed by them are to be carried out by the FPO to some extent. However, it needs strong foundation. The farmers can't easily translate value chain. It involves investments, experience and adequate time. The necessary investment to transform primary production to value added products need to be generated from both public and private sectors. Several FPOs are formed with small groups. They may find it difficult to raise the required funds and which may find it difficult to get the economies of scale. Consequently, there are chances of failure to provide significant return to the members in due course of time. Thus, it requires convergence of several stakeholders including all the sectors such as public, private, and NGOs to empower FPOs toward sustainability. However, the state government, NABARD, SERP, SFAC, and private groups are moving in right direction. to create FPOs in each district. This may help in long-term sustainability for the value of the farmers. There is high probability of duplicity of efforts in the same district and for same commodity groups. It is possible that each agency is not completely aware of the efforts by other agencies in the same geographical region. These challenges need to be addressed through innovations in product markets and institutions for long-term sustainability.

Production conditions and agricultural markets play an important role in contributing to farmers' income. Production is organized through technological innovations, timely supply and efficient use of inputs and operations. On marketing front, several innovations exist to improve agricultural markets including contract farming, private markets, direct purchase centers, farmer markets, and farmer producer organizations. Use of warehousing receipts as collateral instruments, ICTs for market information, single license for operation of traders, establishment of terminal markets, and futures trading play important role for effective functioning of agricultural marketing system. In addition to these innovations, domestic market reforms are required to make market access more equitable and widespread. Development of cold storage and warehousing facilities can make important contributions to farmers' incomes.

Market linkages through FPO/Farmers' company, Farmers Federation depend on dedicated leadership and group dynamics of the stakeholders as well the support from government. FPOs offer two main advantages. It increases in farmers' buying power and selling power leads to enhanced efficiency in marketing of their produce. Simultaneously, farmers can reduce their costs and risks and increase their profits, i.e. net income. Thus it is an answer to the twin problems of production and marketing by converging the production and marketing processes and functions of various stake holders. Andhra Pradesh government is encouraging the concept of Farmers Producer Organizations, FPOs on a large scale to help farmers to reduce the cost and improve farm productivity and profitability. Horticulture and aquaculture is catching up in Andhra Pradesh. Dairy development is on priority. Non-farm income sources through marketing offers great scope in the context of gaining momentum for FPOs . There is ample scope for innovation of institutions. Therefore, the present paper attempts to understand the concept, approaches, building and linking of FPOs to production and marketing for income generation in agriculture sector of Andhra Pradesh.

The concept of institutions

We need to understand the basics of institutions for creating the new institutions or improving the existing ones. Institutions are the 'going concerns' that order the relationships among individuals in the society. It includes laws, constitutions, moral and ethical structures and customary and accepted ways of doing things. The market itself is an institution as idealized by economic theorists or implemented in the practice of business. Institutions control, restrain or at least influence almost every activity. It involves the interpersonal relationships in a complex modern society. It must enhance protection from the harm that others may do them. It defines the rules of the game and there by define the structure. Continuity plays a crucial but dedicated role in the process of change. Change is the law of life and intertwined with moral and ethical values. There is a great scope and potential for institutions dealing with production and marketing of agricultural commodities.

Institutions perform three main functions. It channels information about market conditions, goods and participants. It defines and enforces property rights and contracts determining who gets what and when. It increases competition in the markets and thereby decreases the inefficiency in the system. The essential elements are market activities, peoples' participation in markets, and finally build effective institutions. Marketing activities are wide ranging encompasses collection-procurement, grading and standardization, processing-branding, wholesale, retailing, distribution channels, transport, storage and warehousing etc. Peoples' participation in markets/marketing activities is to be ensured. Market supporting institutions (within the given frame work), are judiciary, banks, firms, property rights, traders, money lenders, state government policies are intertwined. Now the question is how to build effective institutions. Institutional design is important. One size does not fit. What can be done in today's world is important. Evolution of institutions takes time. It requires dedicated leadership. Norms, culture, social conditions determine the success. It is a cumulative process and complements the existing institutions.

Key approaches towards institution building

a) Complement what exists to increase the effectiveness. The number of procedures to register a business is rather higher. Corruption in high places is to be curtailed. Over regulation of business activity particularly in agricultural pricing is to be avoided.

b) Innovate to identify institutions that work: It occurs at national level (public policy), private commercial level, communities and civil society level. We need to support private agents, experimentation, pilot projects and collaboration.

c) Connect communities through information flows and trade. For example, the diverse groups of trading partners, foreign players are to be connected to enforce contracts and adopt new technologies

d) Promote competition to create demand for institutional change or improve quality of existing institutions (R&D in public and private sectors)

e) Motivate the farmers toward their felt needs and mobilize them.

Building institutions for linking farmers with markets

As the economy grows, new problems are arising. Therefore, new institutions are required to address the emerging problems. Evolution takes place by bringing the relevant changes in the existing institutions or building new institutions. Government, cooperatives, private traders, public-private partnerships etc. exist. Community actors and players in the international markets such as WTO, WB, FAO, IMF, IFC also support the changes in the arrangement of institutions. The concept is to be motivated among officers participated in areas relevant to them. Young graduates of agribusiness particularly from IIMs, HRM etc could rope to build new organizations to achieve coordination and scaling up to take the advantage of organized retail, food processing industry. Cold storage investment could be left to free enterprise. The government need to provide seed capital and offer other administrative support to these new ventures. Political will is must for second/evergreen green revolution in India

Changes in political and social forces necessitate the formation of new institutions. Policy makers also create new institutions rather than modify the existing ones. Strong local governments can support regulatory agencies at the local level. The key is to identify what type of institution is needed rather than to assume that a particular structure is needed. There is need share the information among stakeholders. For example, bankers lack information on credit worthy nature of potential borrowers. Everybody's property rights and contracts are to be clearly defined and enforced including tenant farmer. Safeguards are required on intellectual property to create healthy competition. Once the institutional gap is identified the next step is to design the appropriate institution. Both supply and demand factors are important. To be effective, an institution must be designed so that the incentives of market actors are aligned to achieve the desired outcome.

Institutional Arrangement: Industry – farmer partnerships and FPO/FC

Since 1990s, there has been commendable growth in corporate and service sector. To sustain growth, there is need to boost agriculture sector particularly rain fed farming. It requires evolving a mutually beneficial model for farmer-industry partnership and providing durable arrangement for production and marketing. The models that are adopted should be transparent to the conditions of the farmer. Arrangement of partnership may vary depending upon local conditions. It could be on lines of cooperative farming or joint ventures or any other suitable model. With little initiative, agriculture can be transformed into lucrative private enterprise. Successful experiments would encourage other farmers to join hands or replicating the model withappropriate modifications. There is great scope for more crop/ product specific, multi crop andcommodity specific models (AMUL type) to be evolved.

Corporate Affairs Ministry, Indian Institute of Corporate Affairs with National Rain fed Area Authority has developed "Innovative Business Models" for rain fed area development. Farmers are to be linked through processing industries for value addition and branding. Currently only 6% of the produce is processed. Agro based industries would turns out to be propellers of rural economic prosperity. Towards evergreen revolution more open market system through amendments in APMC Acts need to be farmer centric, industry- driven (organized retailing) and knowledge-based strategy are essential. The concept of community based organizations (CBOs) for small farmers will be the best solution to minimize the risk of price fluctuations. New initiatives with innovative ideas through village panchayats, NGOs with the support of government and people' participation can only solve the problem to some extent backed by strong political will to make a country farmer friendly and hunger free.

Farmers' company

Contact farming works in bits and prices, Organic farming and cooperative farming have several limitations. There are success case studies in few areas. In the supply chain either large business must procure directly from farmers or must get closer to the consumer. Mere opening up the retail sector to FDI will not bring about the desired changes. FDI potential foreign entrants in Indian multi - brand retail sector requires to source at least 60% of produce from small farmers having land holdings. More effective interaction with organized marketing networks calls for scale and coordination. It requires organization of farmers into larger production units, cooperatives or farmer companies. Government help is required to create new societies or companies. The key is organization and enterprise nature of the farmers. It is to be registered under company's acts. Governing body consists of farmers' representatives, corporate civil servants. Paid professionals having knowledge in agriculture, and RD do management. The membership includes farmers, SHGs, consumers and cooperatives. Funds are raised from different sources. Regular monitoring and evaluation at various levels will help in improving the performance of an FPO. It includes regular interaction with FPOs through district level implementing unit and experts and consultants. To maintain accountability and transparency of the FPO formation and performance, a third-party auditing will be encouraged. The performance evaluation will be done with regular interval to track the changes with its functioning and processes. This feedback helps to improve the forth coming FPOs. The potential areas for FPOs in Andhra Pradesh in general fall under the following categories.

- System based FPOs: Organic farming, Natural farming, IFS etc
- Community based FPOs: Tribal agriculture- Minor forest products like honey
- Gender based FPOs: Women/Men.
- Commodity based FPOs: Crops fruits, spices vegetables and flowers, dairy (The most important commodities are paddy, groundnut, corn, coarse cereals, Jack fruit, coconut, coffee, pepper, turmeric, milk, fish, prawns etc)
- Agro-processing: Value added products.
- Input based FPOs: Seed production, input outlets
- Bio-resources: Biofuels, vermin-compost.
- Renewable energy: Solar and wind energy.
- Certification: Medicinal plants and products, Organic products etc.

There is a need to establish few pilot projects for crops, horticulture, dairy, aquaculture, retailing, IFS, organic farming etc. in vital farm value chain segments as scalable models of sustainable production systems for its wider applications. This will create the knowledge base on the ways to create sustainable production systems and disseminate it across key stake holders. The strategy of pilot interventions of the creation of farmer groups/organizations to establish sustainability and its long run applications should involve all participating departments and other agencies including departments of agriculture, horticulture, sericulture, animal husbandry, fisheries, groundwater, marketing, forestry, civil supplies and universities, Acharya N G Ranga Agricultural University, S.V.Veterinary University, Dr YSR Horticultural University. ATMA and KVKs can expand market apart from technology. A cluster of villages in each district needs to be identified for pilot FPOs.

SFBC-State Level Federations of FPOs to create a State level umbrella support for the member FPOs. A state level company is to be established as in the case of other states.

The Problem of Convergence

The idea of FPO is novel and ambitious initiatives. It requires convergence of several stakeholders including several sectors such as private, NGOs and public including central and state agencies to create FPOs in each district. It is a long run solution. It takes time to become a full pledged institution for delivery of the services. Since it involves long time horizon, the question of sustainability become an important issue. As several departments are involved, particularly from the government, the efforts are either diluted or duplicated. It is possible that each agency is not completely aware of the efforts by other agencies in the same geographical region. On the other side, the convergence problem arises in coordinating various functions and functionaries of the value chain. It terminates with bureaucratic hurdles. If the sustainability is at stake, it fails to provide the expected returns. Several FPOs form with small groups which may find it difficult to get the economies of scale. Relevant business plans, linking FPOs through digital platforms, transfer of technological innovations to facilitate forward and backward linkages including bank credit at district, *mandal* and village level offers for realization of better price discovery. Institutional innovations and building strong linkages will contribute to achieve better performance and growth.

Environmental Markets for Sustainability of Small Farmers

There are several challenges stemming from climate change, particularly extremes of precipitation and drought events which would be expected to increase and thereby the question of sustainability of farming is at stake. It creates hardships to farmers, particularly small and marginal. Moreover, there are several problems associated with environmental markets. However, farmers have options to mitigate and adapt to climate change. In this context the environmental services need to be internalized for the benefit of farmer producers. It could create a meaningful new income streams for farmers. The value chains must compensate the producers for mitigating the negative impacts of climate change. Standard criteria need to be evolved in order for farmers to receive value for ecosystem services such as the practices undertaken to protect the environment. In this context, we need to establish vibrant markets in which farmers can maximize value. Several principles are of particular interest to producers to operate in a transparent, accountable process with robust input from stakeholders in order to make these programs more effective.

Bio-resource technologies are offering tremendous benefits for environmental improvement and providing sustainability to farming communities. Renewable energy production such as biodiesel, biogas, solar and wind power etc., provide solutions to each of these issues and significant benefits for consumers and the environment. We have to advocate for farmers to add value to agriculture while consumers have better choice due to greater market competition. Therefore, we need to evolve policies that encourage expanded markets for these energy sources. Selection and implementation of practices that address specific conservation concerns, such as soil, water and biodiversity is also necessary. We have to encourage producers and evaluate to identify their natural resources concerns. Producers in the market for land could even ask existing landowners to have a resource stewardship evaluation to use the results in land acquisition decisions.

Market innovations for sustainable agriculture

Innovations in market institutions that encourage sustainable agriculture practices are necessary. FPO concept is a relevant platform and it is an opportunity particularly small holders. Small-scale initiatives that use sustainable production practices are supported by market demand, and create innovations in the institutions that govern sustainable practices and market exchanges. Market responds to both local and distant consumers' concerns on the quality of the food. The initiatives rely upon social values, health, nutrition, food safety, promotion of youth and livelihoods to adapt sustainable practices to local contexts, while creating new market outlets for food products. Specifically, private sector is leading partnerships with the public sector and NGOs to build market infrastructure, integrate sustainable agriculture and exchange of transparent information about market opportunities.

The system innovations allow new rules for marketing and assuring the quality of products. New forms of organization that permit multiple roles in the food system, farmer, researcher, consumer and intermediary. New forms of markets and new technologies for sustainable agriculture such as effective biopesticides and soil analysis techniques get importance. The public sector plays a key role in providing legitimate political and physical spaces for multiple actors to jointly create and share sustainable agricultural knowledge practices and products. Participatory guarantee systems are one such example in case of smallholders in Indian markets. PGS attempts to create communities of small farm holders with greater access to and control of marketing systems.It works directly with farmers and groups in various villages. As non-profit organizations for farmers it is ingrained in all their activities. This is visible at the front-end marketing sector, which has various mechanisms to ensure that farmers own equity in the businesses assisting in the marketing of organic production. It not only connects farmers to markets in a systematic manner but also helps to create communities at a manageable level by the farmers themselves. A strong movement of participatory and fair market systems has been growing over the years.

Although PGS offers certification to communities, this is then owned by them. There has been multiplicative growth in the PGS farmer group/family base as well as in the markets. PGSOC took the initiative to institutionalize PGS and combine it with smallholder farming systems and markets. This institutional structure assists in the creation of farmer communities, where the collective owns the market supply chain system and certification mechanism. This system did not previously exist in markets and is thus an institutional innovation. The creation of such an innovative system ensures better transparency for consumers. The challenge is to address the growing distance to markets and the associated requirements of transparency and fairness. Using appropriate technological interventions will mitigate these challenges. Collectivizing farmer groups as envisaged by PGS also tends towards fairness, with groups insisting on prices from the markets rather than being left at their mercy. This trend in organic markets is seen even in fair trade systems where over 70 percent of the produce is organic (Singh, 2004). PGS helps and empowers farmers to continue this trend.

Conclusion

Success of FPO/PC depends on specific to the needs of farmers, location, crop, market, supply chains, inputs, infrastructure etc. In Telangana state, Ankapur model is based on seed production, resource conservation, and optimum combination of enterprises in a village. It is organized by village panchayat. Mulkanur is an integrated model based on a registered society offering wide range of services in banking and marketing with an operational area of 14 villages. Vegetable market at Narakoduru is a collection centre for a few villages serving well even with middlemen. However, the farmers market developed at Vedullapally for flowers and vegetables, it is not utilized due to lack of functional body. Dairy based societies such as Sangam, Jersy, Tirumala, Dodla, Heritage are well organized on a large scale with strong network in dairy sector.

Clearly, there is missingpolicy for integrating the needs and services. A model with common structure can't serve the purpose. Crop specific, product specific models need to be designed. FPO is expected to bridge the gap in the system. Further, the duplication of efforts by different stakeholders will be corrected with convergence approach. The success of these models is attributed to dedicated leadership coupled with farmers' participation. The institution of SHG model is highly successful model in Andhra Pradesh. The spirit of these institutions needs to be emulated for successful operation of FPOs. Linking SHGs to FPOs in specified areas enhances the efficient functioning of FPOs. Scale can be achieved by offering the required technological inputs for production and doing sorting, grading, packing, and reduced intermediaries apart from financing enhance in marketing, will certainly contribute more than 20% of higher income to farmers. Sustainability of FPO is not an easy task. It requires dedicated leadership. Further sustainability depends on the distributional gains of the members of the organization, which in turn depends on the functioning over time. Since it enters into the business and vested with paid management guided by few individuals, transparency and accountability play important role. Moreover, farmers' participation over a time is also important as the preferences and opportunities are changing fast. As farmers sell everything in wholesale and buys everything in retail, the concept of FPOs makes them to reverse the situation by joining their hands to become entrepreneurs. Farmers sell everything in wholesale and buy everything in retail. As cluster is to be homogeneous group of farmers, cooperation forms the basis the cluster approach pays the dividends.

References

1. FAO. 2012.*Good Practices in Building Innovative Rural Institutions to Increase Food Security.*
2. GoAP. 2016.*Rythu Kosam: Andhra Pradesh Farmer Producer Organisations Promotion Policy –2016.* Operational Guidelines GoAP and ICRISAT.
3. Ashish Gupta. 2017.*Participatory guarantee systems: The case of smallholders in Indian markets in Innovative markets for sustainable agriculture How innovations in market institutions encourage sustainable agriculture in developing countries.* Allison Loconto Anne Sophie Poisot Pilar Santacoloma (Edited) Food And Agriculture Organization of the United Nations (FAO) and Institut National de la Recherche Agronomique (INRA) Rome, 2016.
4. NCU-Bengaluru. 2017. National Consultation on Farmer Organisations: Status and Prospects, 25-26 July 2017; ICAR-NIANP, Bengaluru.
5. Raju KV, Kumar R, Vikraman S, Moses Shyam D, Srikanth R, Kumara Charyulu D and Wani SP. 2017. *Farmer Producer Organization in Andhra Pradesh: A Scoping Study. Rythu Kosam Project. Research Report IDC-16.* International Crops Research Institute for the Semi-Arid Tropics. 160 pp. ISBN 978-92-9066-592-2.http://oar.icrisat.org/9870
6. Reddy, R. 2017. *Brief Note on the working of the society:The Mulkanoor Cooperative Rural Bank and Marketing Society Ltd.*, Mulkanoor, Mandal: Bheemadevarapally, District: Karimnagar, Mimeo.
7. T. Shaw. 2016.*Making farmers' cooperatives work.* Sage Publishers, New Delhi.
8. The Hindu. 2014.

   http://www.thehindu.com/news/national/telangana/ankapur-a-role-model-for-telangana-villages/article6497232.ecehttp://www.thehindu.com/news/national/andhra-pradesh/seed-bowl-of-telangana-sitting-pretty/article8553744.ece 14.10.2014.
9. The World Bank. 2012.*For agricultural innovation systems.* Washington, DC.

❑❑❑

Chapter 16

Self Help Groups for Dairy Entrepreneurship Development

Subash S, Devi M C A and Dixit P K

ICAR- NDRI, SRS, Bengaluru

Indian agricultural system is predominantly a mixed crop-livestock farming system, with the livestock segment supplementing farm incomes by providing employment, draught animals and manure. India has the highest livestock population in the world with 50% of the buffaloes and 20% of the world's cattle population, most of which are milch cows and milch buffaloes. India's dairy industry is considered as one of the most successful development in the post-Independence period. India ranks first in milk production, accounting for 18.5% of world production, achieving an annual output of 155 million tones during 2016 with a growth rate of around 4.6%. Further, per capita consumption of milk in the country was 340 g a day as against 299 g globally. Livestock sector plays a cardinal role in the Indian economy, particularly in livelihood of the rural population. About 70 percent of rural households' own livestock and interestingly, it reflects plurality of small and marginal farmers. Furthermore, women form the primary work force for livestock production besides being their direct involvement in most of the operations. Empowerment of rural women could be well achieved through developing the entrepreneurial skills and their financial independence through Self Help Group (SHG). Development of entrepreneurship is crucial in harnessing untapped human resources. Hence, SHGs served as better platform to empower farm women through dairy entrepreneurship. The fact that dairying could play a more constructive role in promoting rural welfare and reducing poverty by generating employment at farm level through SHGs is increasingly being recognized.

Self Help Groups: A Harbinger of Change

Self-initiated groups are organizations formed by people with a common problem or situation, for the purpose of pooling resources, gathering information and offering mutual support for mutual benefit. Farmers' groups of all types have an important role in agricultural and rural development. They provide space for participation, which contributes to group members' ownership of the issue at hand as well as its solutions. This in turn builds group cohesiveness, solidarity and promotes mutual support. They can be the platform for building a sense of community, a social support system, increasing self-confidence, learning together and providing a sense of equality. A group can be taken seriously in a wider environment. Farmers' groups help empower individuals as well as the groups themselves. SHGs play a significant role in economic and social empowerment of farmers through training and capacity building by involving themselves into direct income raising, production facilitating, cost saving and community development activities.

The concept of self-help is not new in India. Rao (2002) pointed out that SHGs were popular even during traditional and ancient times, and were involved in the main functioning of society. Right from the colonial era, policies favoring the promotion of cooperatives to protect the over-reliance of farmers on money lenders were promoted by the government. Following independence, several attempts were made by the government to enhance local participation. Cooperative farming with local participation, for instance, was the focal point of the government's second five-year plan while the seventh five-year plan (1985–90) promoted schemes favoring participation of people at the grassroots level (NABARD, 2010). Some government agencies, now involved in the development and promotion of SHGs in India, include the National Bank of Agricultural and Rural Development (NABARD), Agricultural Technology Management Agency (ATMA), District Rural Development Agency (DRDA), National Horticultural Mission, Cooperative Department, Khadi and Village Gram Udyog Commission and Krishi Vigyan Kendra (KVK)-Farm Science Centre. Besides government efforts, a few NGOs are also involved in the promotion of SHGs.

The concept of promoting rural institutions through the provision of training and finance was first adopted by an NGO in southern India - The Mysore Resettlement and Development Agency (MYRADA). The success of MYRADA was closely followed by NABARD in 1985. In 1992, NABARD adopted the SHG bank linkage approach as the core strategy for the banking system to increase their outreach to the poor with 500 SHGs. Usually, membership did not exceed twenty in any one SHG (Fernandez, 2006). In India, the majority (90%) of SHGs consist of women members only (NABARD 2010). The biggest revolution in microfinance started in the late 1990s and early 2000s with the involvement of both national and international agencies. Under the SHG bank linkage approach initiated for development of rural sector including agriculture, four distinct models are adopted by a bank to finance SHGs: (1) Banks provide microfinance to NGOs for lending to SHGs and ultimately to the micro entrepreneur: this model caters for approximately 27% of SHGs; (2) Banks provide direct financing to SHGs for on-lending to micro entrepreneurs: this caters for approximately 17% of SHGs; (3) Bank finance directly to SHGs for on-lending to micro entrepreneurs with the intervention of NGOs as social mobilisers and facilitators: this covers 56% of SHGs; (4) Bank loans directly to individual members of SHGs upon recommendation by SHGs and NGOs. In this case, the NGO assists the bank to monitor, supervise and recover loans. The group uses credit from the bank for economic improvement by investing in group projects or inter-loaning to other farmers, the latter being the more common practice.

Dairying in Entrepreneurship Development through SHGs

Dairy farming plays significant role in sustaining the rural livelihoods, although the phenomenon of farmers' suicides, migration, malnutrition /ill health are widely prevalent in rural India. The Sustenance of rural livelihoods is currently at stake than ever before, in the face of economic liberalization. Livelihoods options are shrinking in rural areas in general and more so in eco-fragile regions, such as drought, desert prone, hilly areas and other under developed / backward districts. Rapidly growing markets for livestock products in general, and dairy products in particular are opening new avenues for enhancing rural incomes. The changing global environment raises questions about the ability of traditional, small-scale businesses in rural areas to share the potential benefits offered by the changing environment. The rapid population growth, coupled with even faster urbanization, creates increasing demands. In India, urban populations in general grow about twice as fast as the overall total, and by 2020 they may exceed the size of rural populations. Such a major demographic trend challenges the capacities of some traditional small-scale businesses to cope with the increasing demands.The growth potential of dairy sector is enormous and it is expected that the consumption of the value-added dairy and food products would grow at a very fast pace. This growth could bring immense benefits to the economy, raising agricultural yields, enhancing productivity, creating employment opportunities through new agro-enterprises and raising the economic standard of a large number of populations. Economic liberalization and rising consumer prosperity is opening up new opportunities for diversification in dairy and food processing sector. Liberalization of world trade will open up new vistas for growth. In this scenario entrepreneurship development in dairy and food sector will be a key driver for promoting & sustaining the momentum of growth. Governments, Institutions and individuals have increasingly visualized entrepreneurship as a strategic intervention for accelerating the pace of development in any economy. Entrepreneurship development is essential for exploiting full potential of dairy and food industry on one hand and providing job opportunities to the teaming population of the country on the other.

As a step in this direction, an action research was conducted in Bangalore Rural district of Karnataka state to identify the effectiveness of women's groups in improving their livelihood through dairying. Development of entrepreneurship through women Self Help Groups (SHGs) exclusively in dairying was taken up with a specific objective of improving the livelihood of SHG members through dairy based activity. Two women SHGs on dairying were formed and trained in scientific dairy farming practices through institute interventions. The two groups viz.,Bhuvaneshawari SHG, at Suradenapura village and Sri Sakthi SHG at Hannumanahalli village were formed and developed in association with local NGOs viz, Peoples Trust and Samastha Micro Finance respectively. The SHG members at the initial stage were assisted in their group dynamics, functioning and securing a loan from a local Grameen Bank for purchase of a Cow for each member. The two SHG groups were motivated to take up the dairying as their income generating activity and through which they could secure increased monthly income of SHG members on an average of Rs. 6000 to 7000. The exposures of members of dairy SHGs to various training programmes and latest technologies have significantly increased their level of knowledge on improved dairy farming practices and as a result, improved in milk production of their dairy animals. The training on processing of various dairy based products is also being given to take them to the next step that is from production to processing. However, the motivating the group members towards taking up the dairying as a entrepreneurship activity was very challenging due to lack of need based entrepreneurship development programmes, marketing avenues for the products, accessibility to technical know-how and availability of quality raw materials etc. Further, the study also reveals that dairy based SHGs besides contributing towards economic development these groups are also important for reinforcing their social affinity. Hence SHG model of technology dissemination in various agro-based enterprises need to be developed for not only production aspects but also processing of the products in order to harness the fullest benefit from any agri-business enterprise.

The major critical determinants for the success of this developed groups includes; Strong cohesive and close-knit group members with adequate support from development agencies; properly trained about the Group functioning and activities; a strong network with developmental agencies including NGOs and local Governmental organizations; Ease in securing the credit facilities with the support of local NGOs; Adequate training support on improved dairying practices and Constant co-operation and support from the different stakeholders.

The constraint analysis revealed the important constraints faced by the SHG members in taking up any income generating activity in general and dairying in particular. The five major constraints includes; Lack of basic infrastructure for the selected income generating activities, In adequate training/ Entrepreneurship Development Programmes (EDP) for SHG members, Lack of exclusive market facilities for SHGs products, Low level of knowledge/skill of SHG members and Insufficient incentives from the government for SHG members. Through adequately addressing these issues by the concerned organizations, the movement of group could usher considerable changes in socio-economic status of the marginalized and poor especially among rural women in our society thereby leading to real women empowerment through dairying.

Key learnings and Strategies to promote SHG based Dairying

Based on the experiences gained during the implementation of this project and also feedback received from respondent-SHG members, suitable strategies are hereby recommended for its implementation by the concerned agencies so as to demonstrate and exploit the potential of SHGs as most viable & enabling platform for taking up the dairying activity as successful income generating enterprise. The strategies include;

- Microcredit interest rate be reduced for dairying
- Providing subsidy loan for women SHGs especially for dairying based activities
- Providing exclusive marketing facilities to the SHG products
- Special packages/incentives for dairy based SHGs from Dairy Cooperatives
- SHG focused approach for distribution of free schemes/incentives
- Focused extension efforts/Entrepreneurship Development Programmes for SHGs by State development departments
- Creation of common infrastructure facilities for SHG members
- Flexibility in loan repayment

Conclusion

The success of dairy-enterprise depends on factors, such as, an efficient yet economical procurement network, hygienic and cost-effective processing facilities and innovativeness in market place. To transform the current status from negative contribution to profitability, particularly to benefit the weaker sections of the society, there is a need for a drastic change in the livestock husbandry policy and greater involvement of the private sector and local communities in development of entrepreneurial avenues and business opportunities in dairying sector. SHG is a movement for women empowerment; dairying through SHG movement plays a significant role in sustaining the livelihood of members. The major challenge in developing entrepreneurship includes social, financial, poor infrastructure, technological and policy challenges. The livestock sector should be treated at par with the industry for availing finance and tax benefits and the Government should encourage Public-Private Partnership to attract larger investments, modern technologies and effective networking for forward and backward integration for the sustainable development of dairying sector.

References

1. Fernandez, A. P. (2006). History and spread of Self -Help Affinity Groups in India The role played by IFAD. Retrieved from http://www.ifad.org/operations/projects/regions/pi/paper/3.pdf
2. NABARD. 2010. Status of micro finance in India 2009-2010, Retrieved from http://www.nabard.org/pdf/Status%20of%20Micro%20Finance%202009-10%20Eng.pdf
3. Rao, V. M. 2002. Women self help groups, profiles from Andhra Pradesh and Karnataka. *Kurukshetra*. 50 (6). Pp. 26-32.

❑❑❑